Math Mammoth
Grade 8 Review Workbook

By Maria Miller

ISBN 978-1-954358-57-7

2024 EDITION

Contents

Introduction

Math Mammoth Grade 8 Review Workbook is intended to give students a thorough review of eighth grade math, following the main areas of grade 8 mathematics based on the Common Core Standards. This means the book includes some topics typically found in algebra 1 courses, plus some geometry and statistics. The book has both topical as well as mixed (spiral) review worksheets, and includes both topical tests and a comprehensive end-of-year test. The tests can also be used as review worksheets, instead of tests.

You can use this workbook for various purposes: for summer math practice, to keep the student from forgetting math skills during other break times, to prepare students who are going into ninth grade or algebra 1, or to give eighth grade students extra practice during the school year.

The topics reviewed in this workbook are:

- exponents, scientific notation, and significant digits
- geometry
- linear equations
- introduction to functions
- graphing linear equations
- irrational numbers and the Pythagorean Theorem
- systems of linear equations
- bivariate data

In addition to the topical reviews and tests, the workbook also contains many cumulative (spiral) review pages.

The content for these is taken from *Math Mammoth Grade 8*.

Please note this book does not contain lessons or instruction for the topics. It is not intended for initial teaching. It also will not work if the student needs to completely re-study these topics (the student has not learned the topics at all). For that purpose, please consider *Math Mammoth Grade 8*, which has all the necessary instruction and lessons.

I wish you success with teaching math!

Maria Miller, the author

Exponents and Scientific Notation Review

1. Find the value of the expressions.

a. $(-2)^4 =$	**b.** $-2^4 =$	**c.** $8^{-2} =$	**d.** $5^2 \cdot 5^8 \cdot 5^{-7} =$
e. $11 \cdot 10^{-2} =$	**f.** $10^3 + 10^4 =$	**g.** $\left(\dfrac{2}{-3}\right)^3 =$	**h.** $\dfrac{12^7}{12^5} =$

2. Write an equivalent expression using the exponent laws, without negative exponents.

a. $(a^{-1})^4 =$	**b.** $(2x)^3 =$	**c.** $(5x)^{-2}$	**d.** $-2s^5t^7t^3 \cdot 4s^8 =$
e. $\dfrac{9a^7}{30a^5} =$	**f.** $\dfrac{x^3}{x^{-2}} =$	**g.** $\left(\dfrac{3x}{-4}\right)^3 =$	**h.** $\left(\dfrac{2a^2}{b}\right)^5 =$

3. Find the value of each unknown.

a. $8^x 8^5 = 8^{24}$	**b.** $(7^8)^{-3} = \dfrac{1}{7^y}$	**c.** $\left(\dfrac{2}{3}\right)^x = \dfrac{16}{81}$	**d.** $\dfrac{(-3)^x}{(-3)^5} = -27$	**e.** $(3^z)^2 = 9^4$

4. Find the true statements.

a. $(a \cdot b)^2 = a^2 \cdot b^2$	**b.** $a^m \cdot a^n = a^{m+n}$	**c.** $(a \cdot b)^{-3} = \dfrac{1}{ab^3}$	**d.** $(a + b)^2 = a^2 + b^2$

5. Natalie wrote the prime factorization of 21,600. Find the error.

$$\begin{aligned} &21{,}600 \\ &= 100 \cdot 216 \\ &= 10^2 \cdot 6^3 \\ &= (2 \cdot 5)^2 \cdot (2 \cdot 3)^3 \\ &= 2^2 \cdot 5^2 \cdot 2^2 \cdot 3^3 \\ &= 2^4 \cdot 3^3 \cdot 5^2 \end{aligned}$$

6. There is one pretty simple number x for which the expression $(x + 1)^3$ is equal to $x^3 + 1$. Find that number.

7. Calculate. Give each answer to a reasonable accuracy.

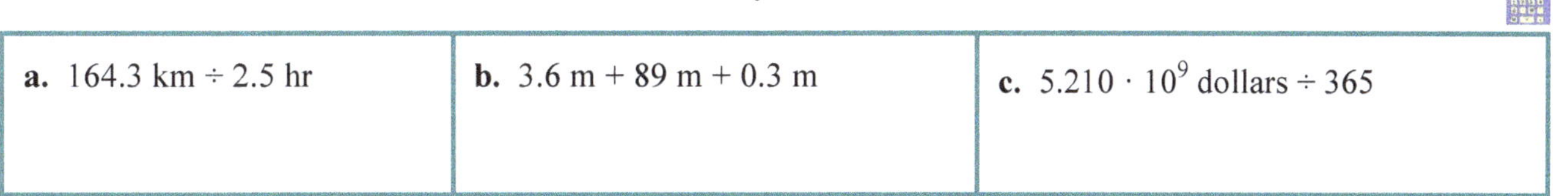

a. 164.3 km ÷ 2.5 hr	**b.** 3.6 m + 89 m + 0.3 m	**c.** $5.210 \cdot 10^9$ dollars ÷ 365

8. Calculate. Give each answer to a reasonable accuracy.

a. Sydney jogs through a 3.5-km jogging track twice every week. What is the total distance she jogs in a year?

b. Your car's odometer shows you traveled 624.6 miles since you last filled your fuel tank. To fill your tank this time, it took 24.3 gallons of gas. What is your gas mileage (in miles per gallon)?

c. What is the area of a rectangle with 2.3 m and 11.9 m sides?

9. Complete the chart by rewriting each distance in scientific notation.

Planet	Average distance from sun (km)	In scientific notation (km)
Mercury	58,000,000	
Jupiter	778,570,000	
Neptune	4,495,000,000	

10. Radioactive elements undergo radioactive decay, where they lose energy by radiation. Half-life is the time required for exactly half of these entities to decay, on average.

Prefix	Meaning
milli	10^{-3}
micro	10^{-6}
nano	10^{-9}

The half-life of thorium-217 is 240 microseconds, and the half-life of uranium-216 is 4.3 milliseconds.

a. Write these two amounts in scientific notation, in *seconds*.

b. How much longer is the half-life of uranium-216 than the half-life of thorium-217? Give your answer in milliseconds.

11. The average speed of a garden snail is about $1.3 \cdot 10^{-2}$ m/s. The average speed of a cheetah is about 27.7 m/s. About how many times faster is the cheetah than the garden snail?

Exponents and Scientific Notation Test

Do not use a calculator for the problems on this page.

1. Find the value of each expression, without a calculator.

a.	b.	c.	d.
$(-9)^2 =$	$5 \cdot 2^{-3} =$	$5^8 \cdot 5^3 \cdot 5^{-9} =$	$\left(\frac{1}{4}\right)^3 =$
$-9^2 =$	$(5 \cdot 2)^3 =$	$(4 \cdot 3)^{-2} =$	$\frac{5^7}{5^9} =$

2. Write an equivalent expression, using the exponent laws, and without leaving negative exponents.

a. $3v^5 \cdot 2^3 \cdot v^3 =$	**b.** $-2b^5a^7 \cdot 5b^3a^8 =$	**c.** $(5x)^2 =$	**d.** $(2x)^{-3}$
e. $2s^{-2} =$	**f.** $(s^{-2})^3 =$	**g.** $\left(\frac{-x^2}{2}\right)^5 =$	**h.** $\frac{25x^9}{15x^4} =$

3. Find the expressions that have the value 3^8.

a. $3^2 \cdot 3^3 \cdot 3^3$	**b.** $\frac{3^{16}}{3^2}$	**c.** 24	**d.** $\frac{3^{11}}{3^3}$	**e.** $\frac{3^7}{1/3}$	**f.** 3^{2^4}

4. Order the numbers from the smallest to the largest.

$2 \cdot 10^8$ $0.4 \cdot 10^6$ $5.2 \cdot 10^7$ $9 \cdot 10^5$ $64 \cdot 10^6$

5. **a.** How many times bigger is $7 \cdot 10^{-9}$ than 10^{-18}?

b. How many times bigger is $2 \cdot 10^6$ than $8 \cdot 10^{-5}$?

You may use a calculator for the problems on this page.

6. A (rectangular) driveway is measured to be 32.5 ft long and 10.5 ft wide.
   Calculate its area, giving the result to a reasonable accuracy.

7. Calculate an estimation for the total cost of providing 12,000 students with a meal costing $2.65 each, five days a week, 36 weeks a year. Assume that 12,000 is accurate to three significant digits, and treat five and 36 as perfectly accurate.

8. The speed of sound in salt water is about 1500 m/s. Calculate the distance sound travels in a year, in salt water. Give your answer in kilometers, in scientific notation, and with two significant digits.

9. The mass of one copper atom is about $1.055 \cdot 10^{-22}$ grams. How many atoms are in 10 grams of copper? Give your answer in scientific notation.

Geometry Review

1. Describe a sequence of transformations that can map figure 1 to figure 2.

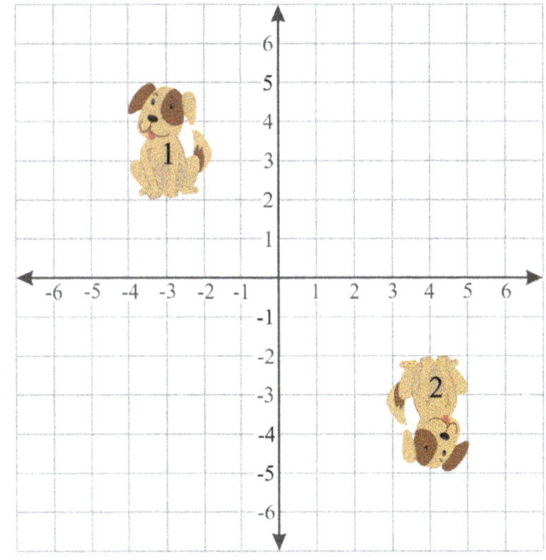

2. Rotate angle DEF both 90 degrees and also 180 degrees clockwise around the origin.

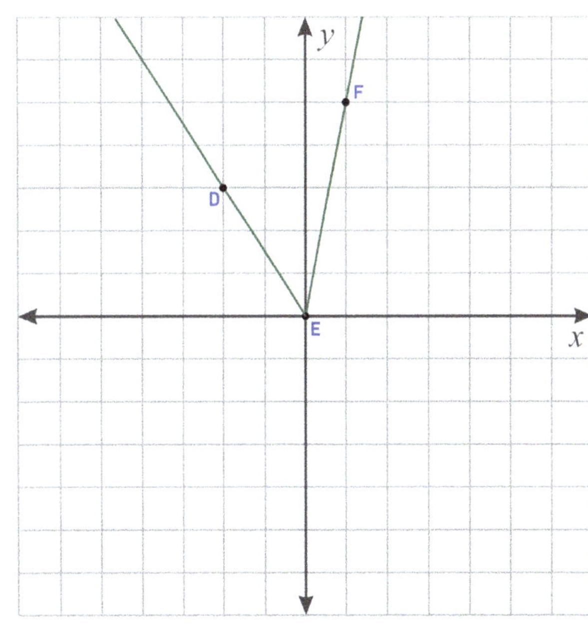

3. A quadrilateral was first reflected in the y-axis, and then rotated around the origin clockwise 90 degrees. Its vertices are now at points $(3, -5)$, $(5, -2)$, $(4, -1)$, and $(1, -4)$. What were the coordinates of its vertices before these transformations?

4. Figure A'B'C'D'E' is a dilation of figure ABCDE with scale factor 3/4. Angle B is a right angle. Check all the statements that are true.

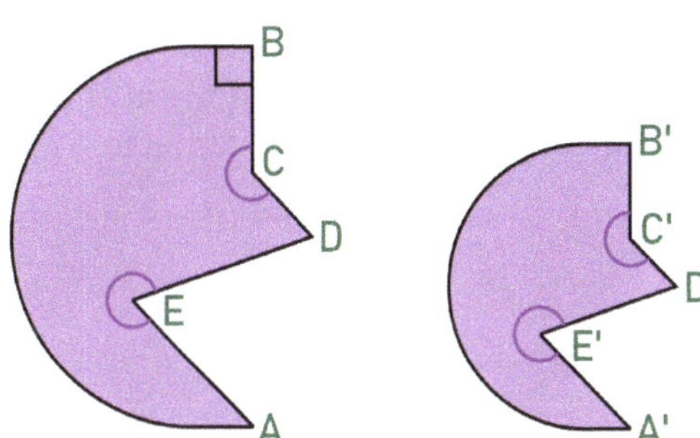

a. Angle B' is a right angle.

b. The measure of ∠CDE is 3/4 of the measure of ∠C'D'E'.

c. ∠E is equal to ∠E'.

d. If CD = 1 inch, then C'D' = 3/4 inch.

e. ∠D is equal to ∠E'.

f. If the perimeter of figure ABCDE is 20 inches, then the perimeter of A'B'C'D'E is 12 inches.

5. **a.** Perform the following sequence of transformations to triangle ABC:

First rotate it counterclockwise around point C 90 degrees.
Then reflect it in the vertical line at $x = -1$.
Lastly, translate it two units to the right and three down.

b. Find another, different sequence of transformations that does the same as the sequence in (a), and starts with a reflection.

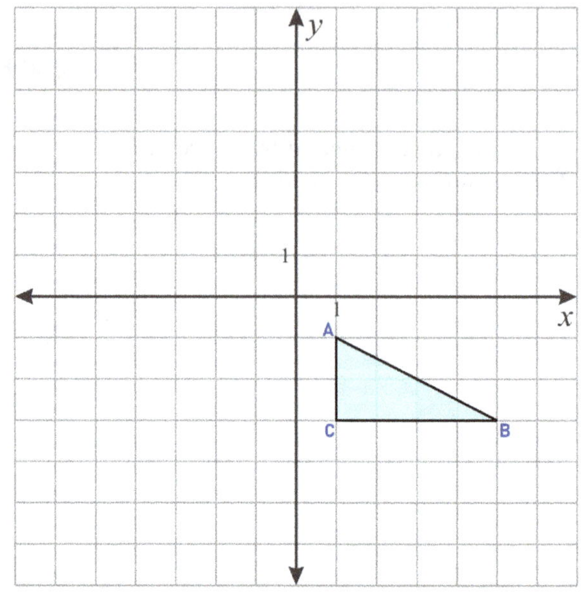

6. Draw a dilation of triangle ABC...

a. from point C and scale factor 1/2.

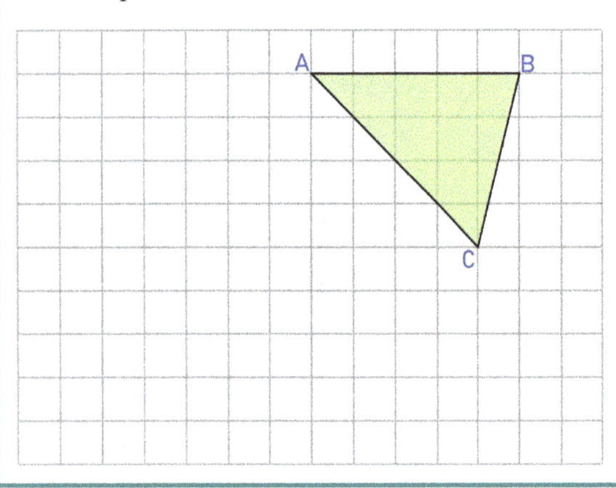

b. from point B and scale factor 2.

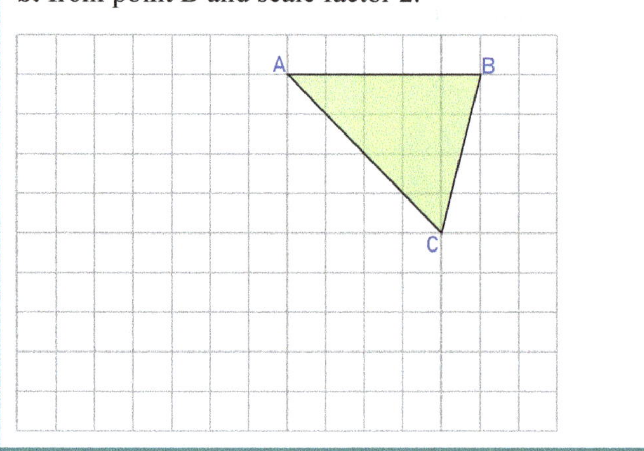

7. Parallelogram EFGH underwent the following transformations :

1. Reflection in the vertical line at $x = -0.5$.

2. Translation 3 units to the left and 4 units down.

3. Dilation from point E'' with scale factor 2.

What are the coordinates of the image of point F after all these transformations?

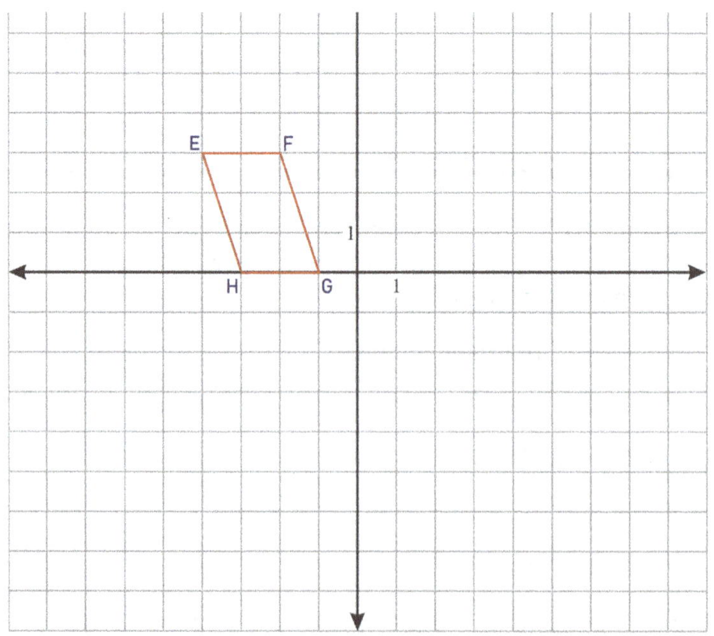

8. Show that the two triangles are similar by describing a sequence of transformations that could map △DEF to the smaller triangle

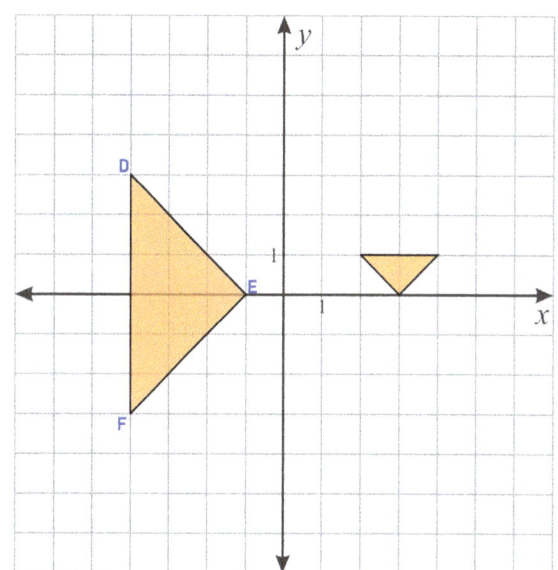

9. Figure PQRS underwent a dilation, then a rotation. Study the coordinates to find out the details about each transformation, then fill in the missing coordinates.

Original figure	Dilation	Rotation
P(−5, 3)	P'(−6, 5)	P"(___ , ___)
Q(0, 3)	Q'(4, 5)	Q"(___ , ___)
R(−1, 1)	R'(___ , ___)	R"(−4 , −5)
S(−4, 1)	S'(−4, 1)	S"(−4, 1)

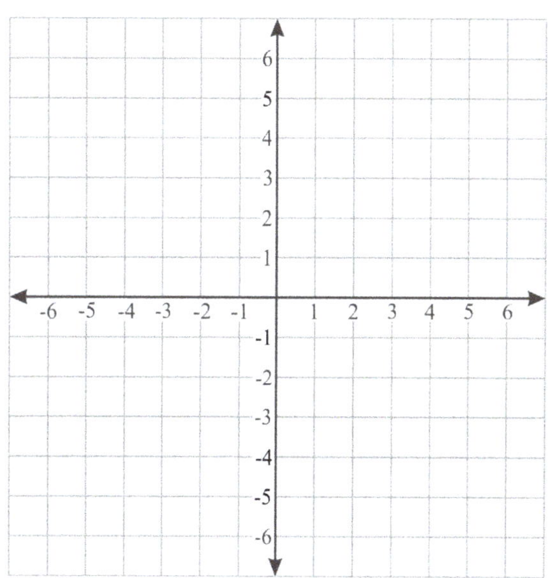

10. **a.** Find the value of x.

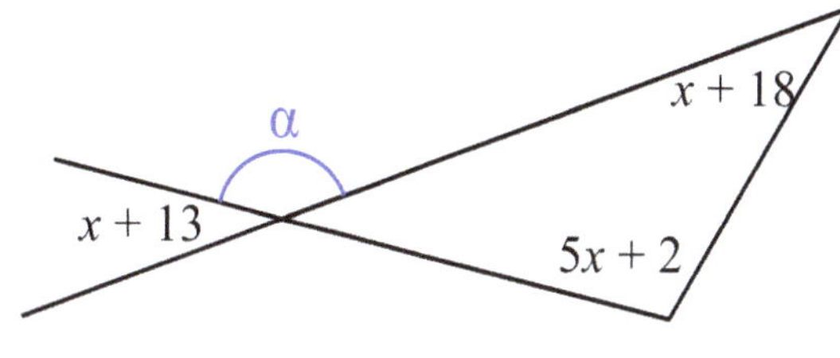

b. Find the value of α.

11. Lines l and m are parallel. Figure out the measure of the angle α. (You may need to mark more angles in the diagram.)

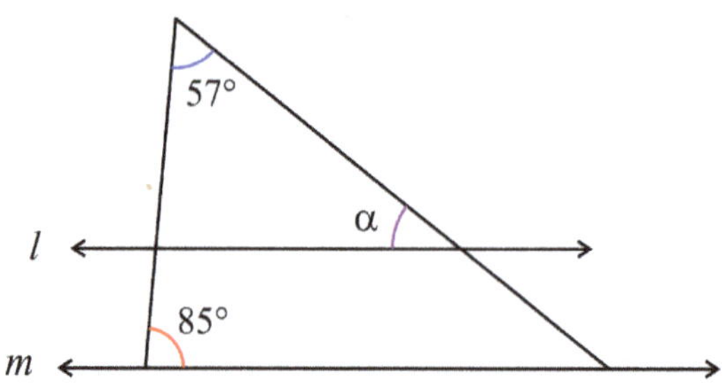

12. Lines m and n are parallel. Find the measure of angle x, and prove why it is what you find it to be. In other words, explain and justify your reasoning. You may need to mark more angles in the diagram.

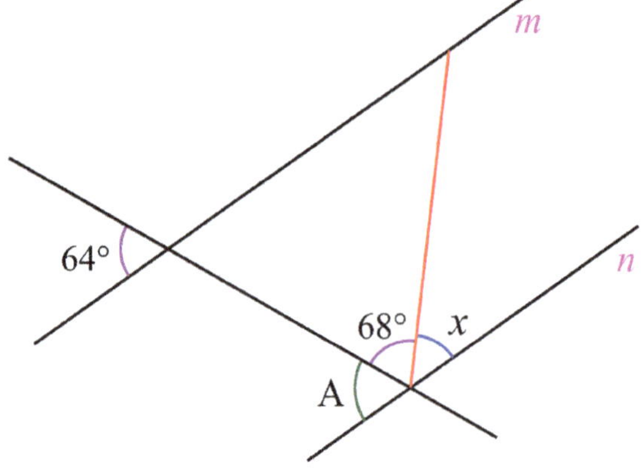

13. A shampoo bottle is in the shape of a circular cylinder. It says it contains 473 ml of shampoo. Its inner diameter is 6.0 cm and its height is 17 cm. What percent of the bottle does the shampoo take up?

14. Compare a sphere with radius 5 cm with a cone with the same radius. What is the height of the cone, given the two have the same volume?

15. Elkhart Ammonium Nitrate Storage in Elkhart, Texas, is a large building consisting of a half sphere on top of a circular cylinder. The stem wall is 36 ft high, and the diameter of the cylinder (which is also the diameter of the sphere) is 115 ft. When the storage is filled with ammonium nitrate, the top part of it (the part inside the half-sphere) forms a cone.

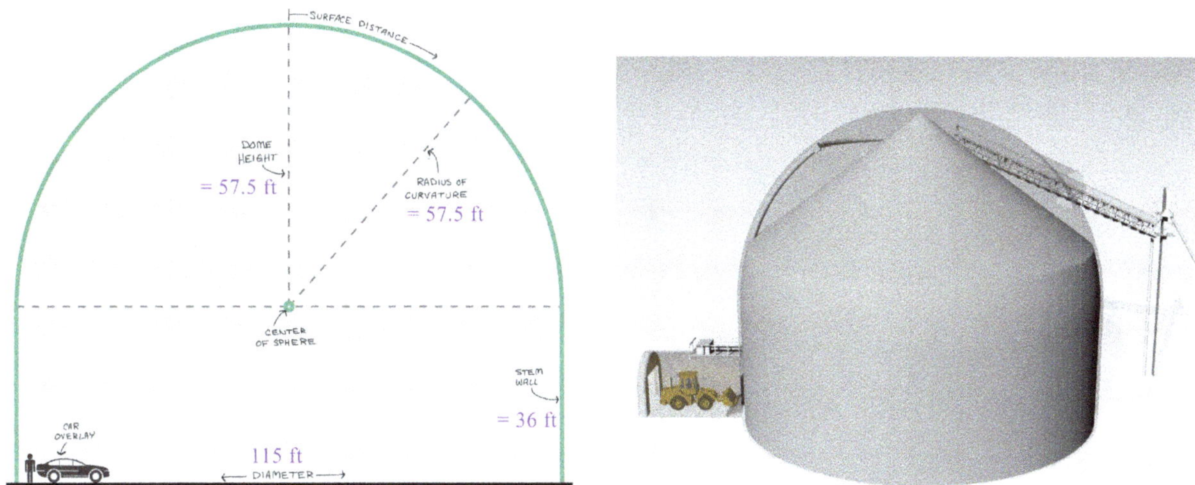

Images courtesy of Monolithic Dome Institute, www.monolithic.com

Find the volume of the ammonium nitrate mound when the cone reaches the top of the structure, to the nearest thousand cubic feet.

Geometry Test

1. Trapezoid ABCD is first rotated 90 degrees counterclockwise around the origin, and then reflected in the horizontal line $y = 1$.

   Draw the image of the trapezoid after these transformations.

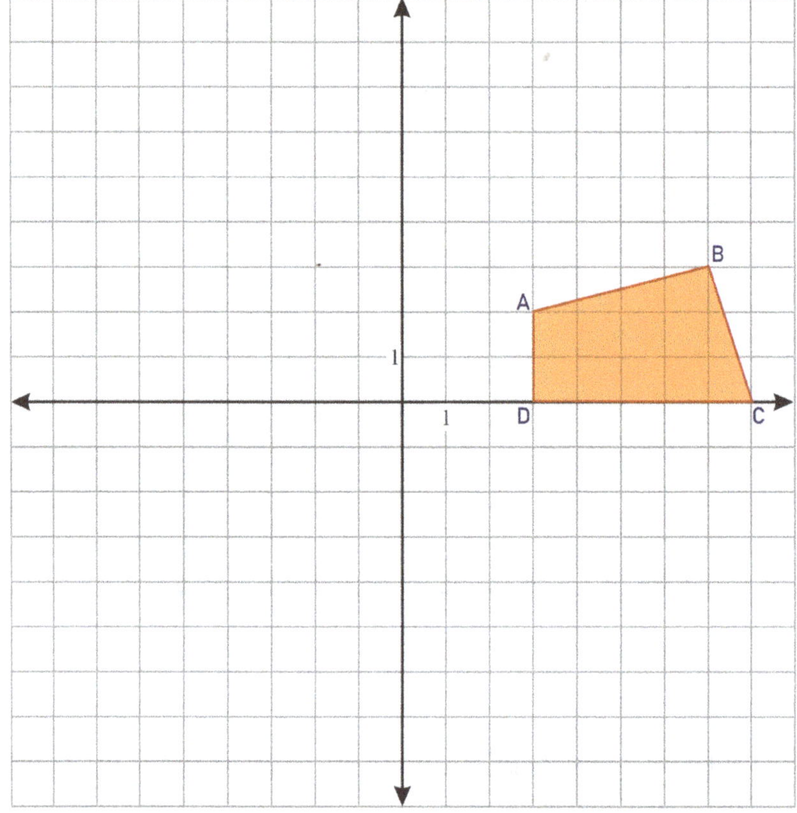

2. Show that the two triangle shapes are similar by describing a sequence of transformations that maps the larger arrow to the smaller arrow.

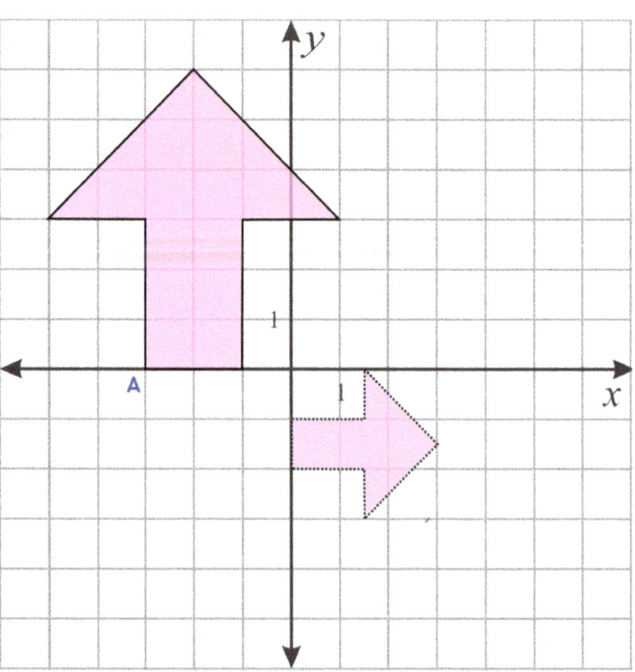

3. A triangle was first reflected in the y-axis, then translated two units down. Now its vertices are at points (−4, −2), (−3, 0), and (−1, −2). What were the coordinates of its vertices before these transformations?

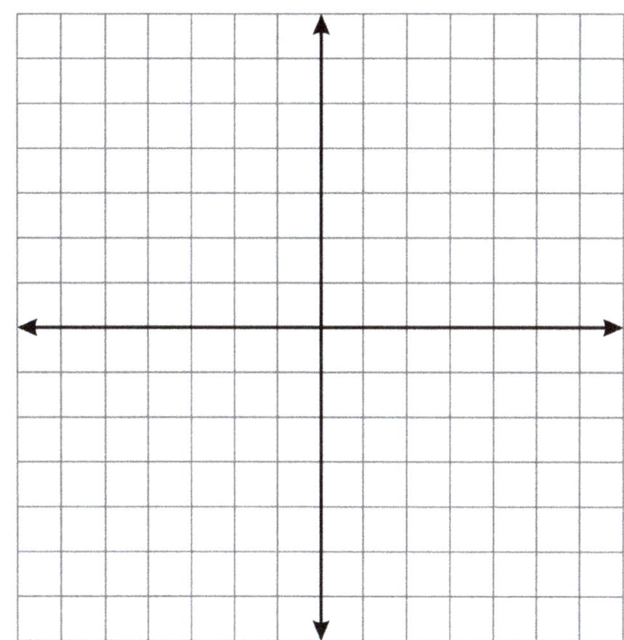

4. A rectangle with vertices at (9, −3), (9, 0), (3, −3) and (3, 0) is dilated with origin as the center of the dilation, and with a scaling factor of 1/3. Then it is translated one unit to the left and one unit down. What are the coordinates of the vertices of the resulting rectangle?

5. Quadrilateral ABCD undergoes a dilation (from 1 to 2) and then a certain congruent transformation, from 2 to 3.

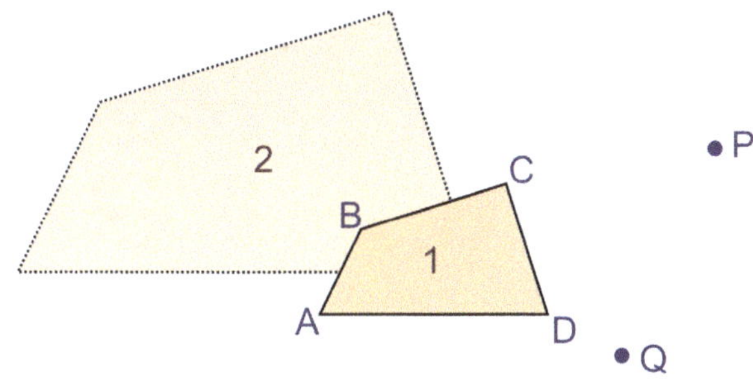

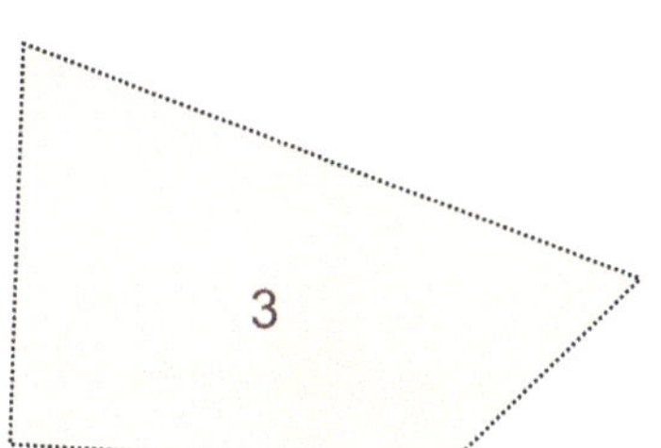

a. Name the congruent transformation.

b. Mark as A' the image of point A under the dilation.

c. Mark as A" the image of point A' under the other transformation.

d. Which attributes of the quadrilateral are preserved in this sequence of transformations? Tick all that apply.

(i) Perimeter

(ii) Area

(iii) Position

(iv) Angle at A

(v) Angle sum

6. Figure ABCD is a parallelogram.
   What is the value of x?

   Explain your reasoning.

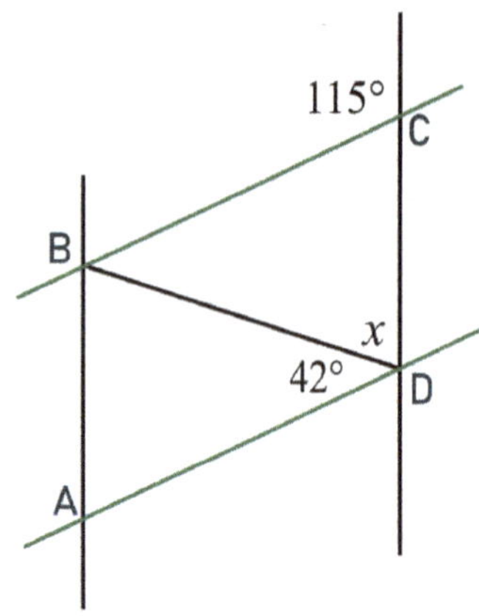

7. Find the measure of angle x.
   Explain your reasoning.

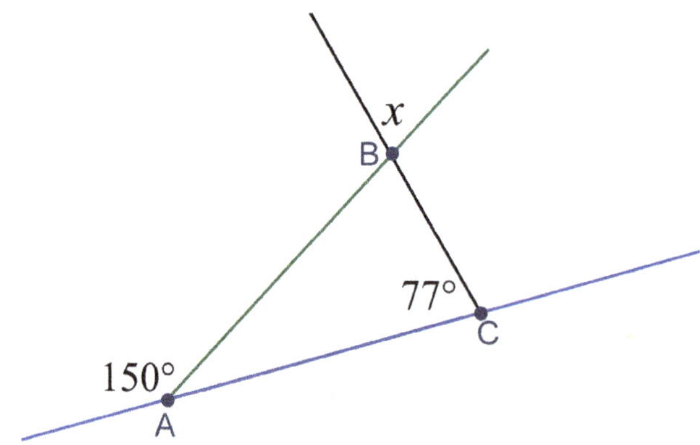

8. Are the two triangles similar?
   Explain how you know.

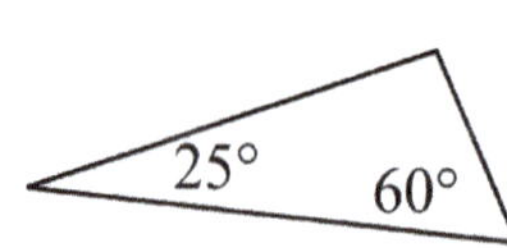

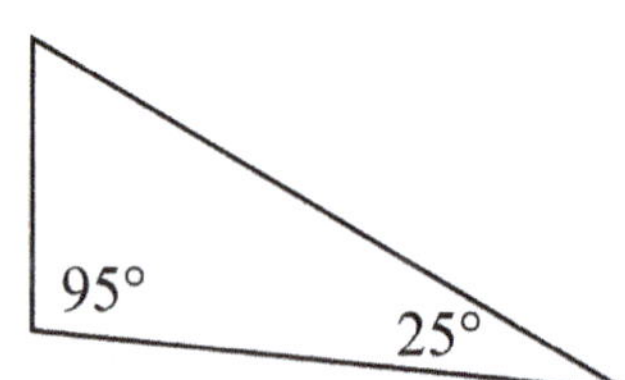

9. **a.** Find the volume of a sphere with a radius of 4.0 cm. Give your answer to a reasonable accuracy, based on significant digits.

b. Find the volume of a cylinder, when its bottom face is a circle with a 4.0-cm radius, and its height is twice the diameter of that circle. Give your answer to a reasonable accuracy, based on significant digits.

10. What fraction is the volume of this cone of the volume of a cylinder it just fits into? In other words, the cylinder and the cone have the same size circle as their base, and have the same height.

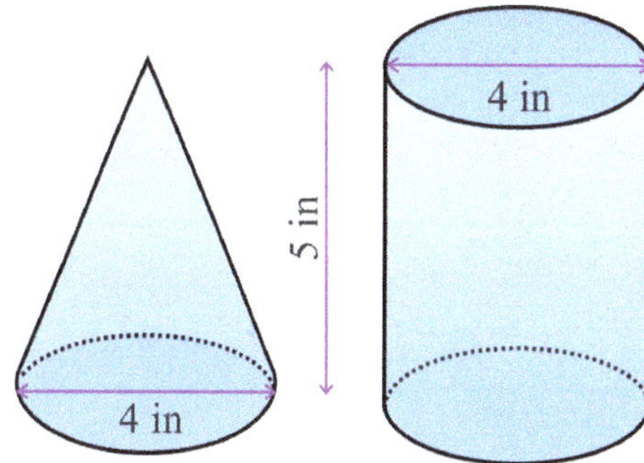

11. Kayla says, "These four limes will give me a half a cup of lime juice." Should you believe Kayla? The limes are spherical, with a diameter of 6 cm, one cup is 240 ml, and 1 ml = 1 cm^3.
Justify your reasoning.

Mixed Review 1

1. Find the value of the expressions.

a. $-4^2 =$	**b.** $5^{-3} =$	**c.** $57 \cdot 10^{-3} =$	**d.** $3^{-6} \cdot 3^9 \cdot 3^{-2} =$
e. $\dfrac{8^6}{8^8} =$	**f.** $\left(\dfrac{-1}{-5}\right)^2 =$	**g.** $2^{-2} \cdot 5 \cdot 2^5 + 4^3 \cdot 4^{-2} =$	

2. Simplify, giving your answer without any exponents.

a. $4 \cdot 10^2 5^2 =$	**b.** $2 \cdot 7^{-1} 7^{-5} 7^3 =$	**c.** $6^9 \cdot 6^{-9} =$

3. Write an equivalent expression using the exponent laws, without using negative exponents.

a. $(2b^5)^3 =$	**b.** $(3y)^{-3} =$	**c.** $6x^{-2}$	**d.** $-2a^3b^7 \cdot 5a^7b^2 =$
e. $\dfrac{x^{-3}}{x^{-2}} =$	**f.** $\dfrac{2a^2}{a^5} =$	**g.** $\left(\dfrac{-2x}{3}\right)^3 =$	**h.** $\left(\dfrac{x}{z^2}\right)^2 =$

4. Compare the numbers, writing < or > in the box.

a. $6 \cdot 10^{-6}$ ☐ $8 \cdot 10^{-7}$ **b.** $5 \cdot 10^{-5}$ ☐ 0.0004 **c.** $2.9 \cdot 10^{-4}$ ☐ 0.0003

5. Anna says that $(5x)^{-2}$ simplifies to $5x^{-2}$. Is she correct? If not, why not?

6. How many significant digits do these numbers or quantities have?

a. 15.0	**b.** 150	**c.** 0.15	**d.** 0.150
e. 4,000 kg	**f.** 4,001 kg	**g.** $7.02 \cdot 10^6$	**h.** $2.060 \cdot 10^{-3}$

7. Rewrite the numbers in scientific notation correctly.

a. $403 \cdot 10^{-3}$

b. $66 \cdot 10^6$

c. $0.2 \cdot 10^{-2}$

d. $0.291 \cdot 10^{-1}$

8. Simplify. Give your answers in decimal notation (not scientific).

a. $2 \cdot 10^{-3} + 5 \cdot 10^{-2}$	**b.** $7 \cdot 10^{-5} + 0.03$
c. $3.2 \cdot 10^{-1} - 0.07$	**d.** $5.4 \cdot 10^{4} - 2{,}000 + 8 \cdot 10^{3}$

9. Calculate with a calculator. Round your answer to the correct amount of significant digits.

a. 0.92 m · 1.3 m	**b.** 814 mi ÷ 32 gal
c. 24,900 mi ÷ 3.14	**d.** 13.7 lb · 4,200
e. $(4.50 \cdot 10^{6}$ dollars) ÷ 13,000 people	**f.** 18 · 35 kg · 115

10. The surface area of the sun is $6.09 \cdot 10^{12}$ km^2, and that is $12 \cdot 10^{3}$ times the surface area of the Earth. Find the surface area of the earth. Use scientific notation.

11. Calculate. Give each answer to a reasonable accuracy.

a. A certain city spent 977.9 million dollars for education in a particular year, and they have 550,000 students (accurate to thousands). How much did they spend per student?

b. What is the area of a triangle with a 56-cm base and 34-cm height?

12. If the distance you need to travel is $9.3 \cdot 10^{7}$ mi (from the earth to the sun), and the time you take to do that is $1.38 \cdot 10^{-1}$ hours, then what is your speed, in miles per hour? (Give your answer in scientific notation.)

Mixed Review 2

1. The points (−5, −1), (−3, 3), and (−1, 0) are the vertices of a triangle. It is rotated 90° clockwise around the origin, and then translated 2 units to the left and 3 units up. What are the coordinates of its vertices now?

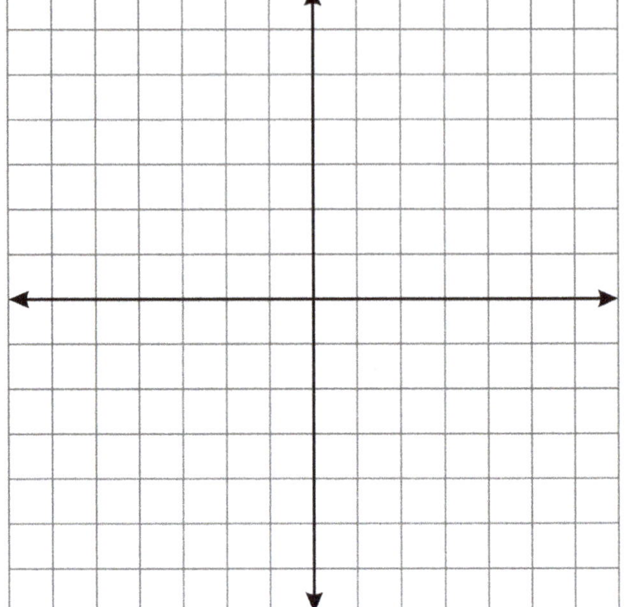

2. A rectangle is first moved 3 units down and 2 to the left, then reflected in the x-axis. Its vertices are now at (12, −6), (12, −10), (9, −6), and (9, −10).

   What were its coordinates before these transformations?

3. Draw dilations of triangle ABC according to the instructions.

a. Draw a dilation of triangle ABC using origin as center, and the scale factor of 1/2.

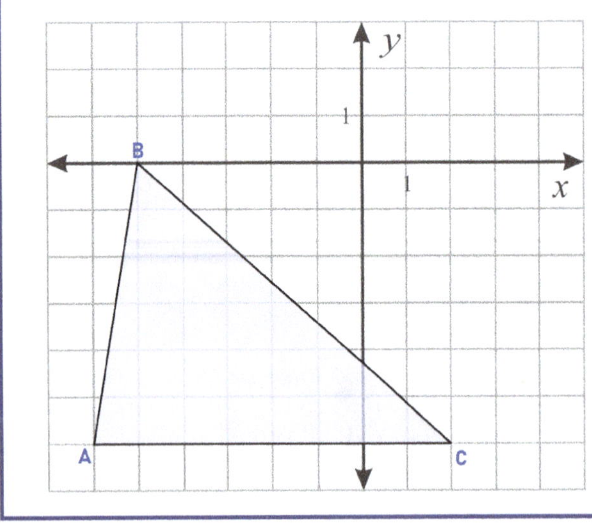

b. Draw a dilation of triangle ABC from point B, again using the scale factor of 1/2.

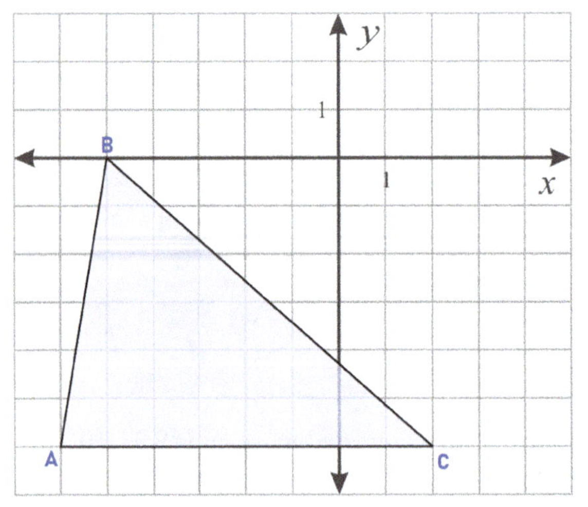

4. Find the value of the expressions. Do not use a calculator.

a. $-2^4 =$	**b.** $(-3 \cdot 2)^{-2} =$	**c.** $3 \cdot 10^{-3} =$	**d.** $5^{-2} \cdot 5^4 \cdot 5^{-3} =$
e. $(3 \cdot 10)^{-3} =$	**f.** $3^2 + 3^3 + 3^4 =$	**g.** $\left(\frac{-5}{6}\right)^2 =$	**h.** $\frac{11^6}{11^8} =$

5. Simplify, and write an equivalent expression that does not have any negative exponents.

a. $(5s^2)^{-1}$ =	**d.** $(xy^{-3})^5$ =	**g.** $(2a)^3 \cdot a^{-2}$ =
b. $(4t^{-3})^{-2}$ =	**e.** $(3xy^3)^{-2}$ =	**h.** $(-10y)^{-3} \cdot (4y)^2$ =
c. $(xy^3)^{-2}$ =	**f.** $(-2w^{-2})^4$ =	**i.** $(-3x)^{-1} \cdot (3x)^3$ =

6. Find the value of each unknown.

a. $(5^x)^2 = 25^4$	**b.** $16^3 = (4^y)^2$	**c.** $(16 \cdot 3^x)^{-2} = \dfrac{1}{6^8}$

7. **a.** Match the expressions and numbers with the same value.

b. Circle the ones that are written in scientific notation correctly.

$0.07 \cdot 10^9$	$7 \cdot 10^4$	$7 \cdot 10^9$
70,000,000	$70 \cdot 10^8$	$0.7 \cdot 10^5$
$0.7 \cdot 10^8$	7,000,000,000	70,000

8. Rewrite in scientific notation correctly.

a. $0.93 \cdot 10^7$ **b.** $23 \cdot 10^6$ **c.** $14.5 \cdot 10^8$

9. A student multiplied two large numbers with a calculator and got this: 2.1E25

a. What does the answer mean?

b. What two numbers could she have multiplied?

10. The population of India is estimated to be $1.5 \cdot 10^9$ in 2030.
That is $2 \cdot 10^4$ times as many people as in Lydia's hometown.
What is the population of Lydia's hometown?

11. Calculate with a calculator. Round your answer to the correct amount of significant digits.

a. 8.04 m · 2.5 m	**b.** 250 · 0.493 kg
c. ($4.71 \cdot 10^4$ dollars) ÷ 1,944 hr	**d.** 946 mi ÷ 26.0 gal

Linear Equations Review

1. Solve.

a. $7x - 3(x - 5) - 2x \;=\; 10$	**b.** $x - \frac{1}{3} = \frac{3}{4}x - 2$
c. $20 - q \;=\; -q + 2(q - 5) - 6q$	**d.** $-52 - 2(x + 14) \;=\; 80 - 11x + x$

2. Heather says, "Five years ago, I was one-third as old as I will be in 45 years." How old is Heather?

3. Solve.

a. $\dfrac{x-1}{2} = \dfrac{3x+2}{7}$	**b.** $x - \dfrac{1-3x}{2} = 5$

4. The density of a liquid is denoted by the Greek letter ρ (rho), and is given by the formula $\rho = \dfrac{m}{V}$, where m is the mass of the liquid and V is its volume.

a. Solve this formula for V.

b. If the mass of a certain liquid is 4.6 kg, and its density is 850 kg/m^3, what volume does it occupy? Give your answer in cubic meters *and* in cubic centimeters.

Hint: To find the conversion factor between cubic meters and cubic centimeters, fill in:

1 m^3 = 1 m × 1 m × 1 m

= 100 cm × ______ cm × ______ cm

= ____________ cm^3

5. Give an example of each type of equation that has $8x - 2$ on the left side of the equation.

a. No solutions:

$8x - 2 =$

b. One solution:

$8x - 2 =$

c. An infinite number of solutions:

$8x - 2 =$

6. Solve. Round the solutions to two decimals.

a. $0.6 + 9.4x - 2 \;=\; x - 4.8x$	**b.** $4 - 0.3(2y + 2.8) \;=\; 1.5(y - 3)$

7. Sarah bought a total of 16 cartons of almond milk, some for the price of \$4.50 and some for \$5.20. Her total came to \$76.20. How many of the cheaper ones did she buy?

8. You're selling homemade bread for \$8.50 per loaf. By how much should you increase the price, so that when you later on have a 25% off sale, the sale price will be \$7.50?

9. The price of an item is reduced by 35%, and then a 7% sales tax is added. You pay \$53.90. What was the original price of the item?

Linear Equations Test

1. Solve.

a. $6 - 3y - 2 + 8y \;=\; 4y + 1 - 9y - 5$	**b.** $12 - (x - 2) \;=\; 10 - 2x$
c. $4x + 20 - x \;=\; 2(x - 5) - 6x$	**d.** $-35 - 3(x - 4) \;=\; 10x + 40 - x + 13x$

2. Ohm's law concerning electric circuits states that $C = \frac{V}{R}$, where C is the current in the circuit, V is the voltage in the circuit, and R is the resistance in the circuit.
Solve this formula for R.

3. Lucas says, "In six years, I will be two-thirds as old as my mom, who is now 66." How old is Lucas?

4. A storekeeper sells a chainsaw for \$288. By how much should he increase the price, so that when he later on has a 20% off sale, the sale price will be \$249?

5. **a.** How many solutions does this equation have?

$$8y + 4 = 4(3 - 2y)$$

b. Modify the equation so that it has an infinite number of solutions.

6. **a.** How many solutions does this equation have?

$$-2(9y - 4) + 8y = 8 - 10y$$

b. Modify the equation so that it has no solutions.

7. Solve.

a. $\dfrac{3x - 5}{4} - 1 = 2x$	**b.** $\dfrac{y - 2}{3} = 3y + \dfrac{5 - y}{2}$

Mixed Review 3

1. Change the base of each power expression.

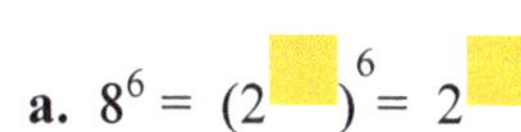

a. $8^6 = (2^{\square})^6 = 2^{\square}$	b. $64^3 = (4^{\square})^{\square} = 4^{\square}$	c. $100^5 = (10^{\square})^{\square} = 10^{\square}$

2. In each case, find the value of the two similar expressions.

a.	b.	c.	d.
$5 \cdot 4^3 =$	$(-5)^3 =$	$-3^{-2} =$	$2 \cdot 5^{-2} =$
$(5 \cdot 4)^3 =$	$-5^3 =$	$(-3)^{-2} =$	$(2 \cdot 5)^{-2} =$

3. The mass of the sun is about $2 \cdot 10^{30}$ kg, and the mass of the earth is about $6 \cdot 10^{24}$ kg. About how many times the mass of the earth is the mass of the sun? Give your answer in scientific notation.

4. Join equivalent expressions with a line.
Some expressions will not be joined.

$\frac{a^2}{b^2}$	$9a^6$	$(3a^3)^2$	$-3a^2$
$3a^6$	$\frac{(3a)^2}{b}$	$\left(\frac{a}{b}\right)^2$	$\frac{3a^2}{b}$
$\frac{9a^2}{b}$	$\frac{a^2}{b}$	$3a^{3^2}$	$3a^9$

5. Find the measure of angle x. Explain your reasoning.

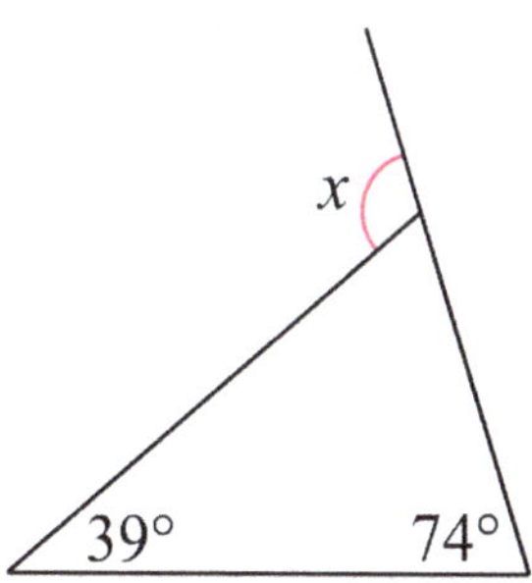

6. Find the volume of this triangular prism, if its height is 36 cm and its base is a right triangle with 10 cm, 15 cm, and 18 cm sides.

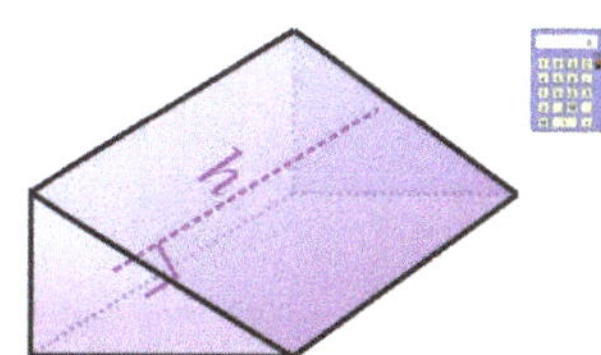

7. Two parallel lines, L_1 and L_2, are cut by a transversal. If angle x is 76°, find the angle measure of y, and explain how you know that.

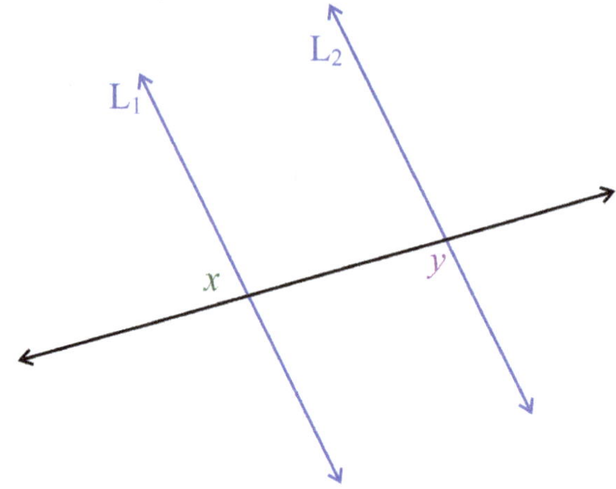

8. A triangle with vertices A(−1, 2), B(−3, 2), and C(−4, 4) is first reflected in the y-axis and then rotated 90° clockwise around the origin. What are the coordinates of the vertices of the resulting triangle?

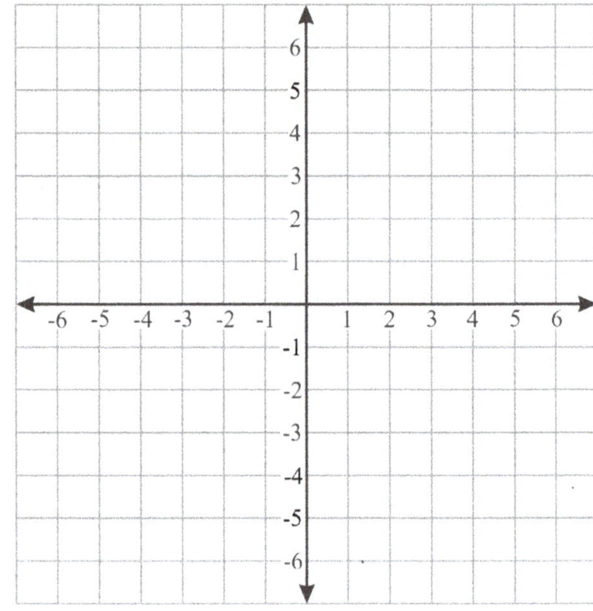

9. Dilate each figure with origin as center and with the given scale factor.

a. scale factor 3

b. scale factor 1/2

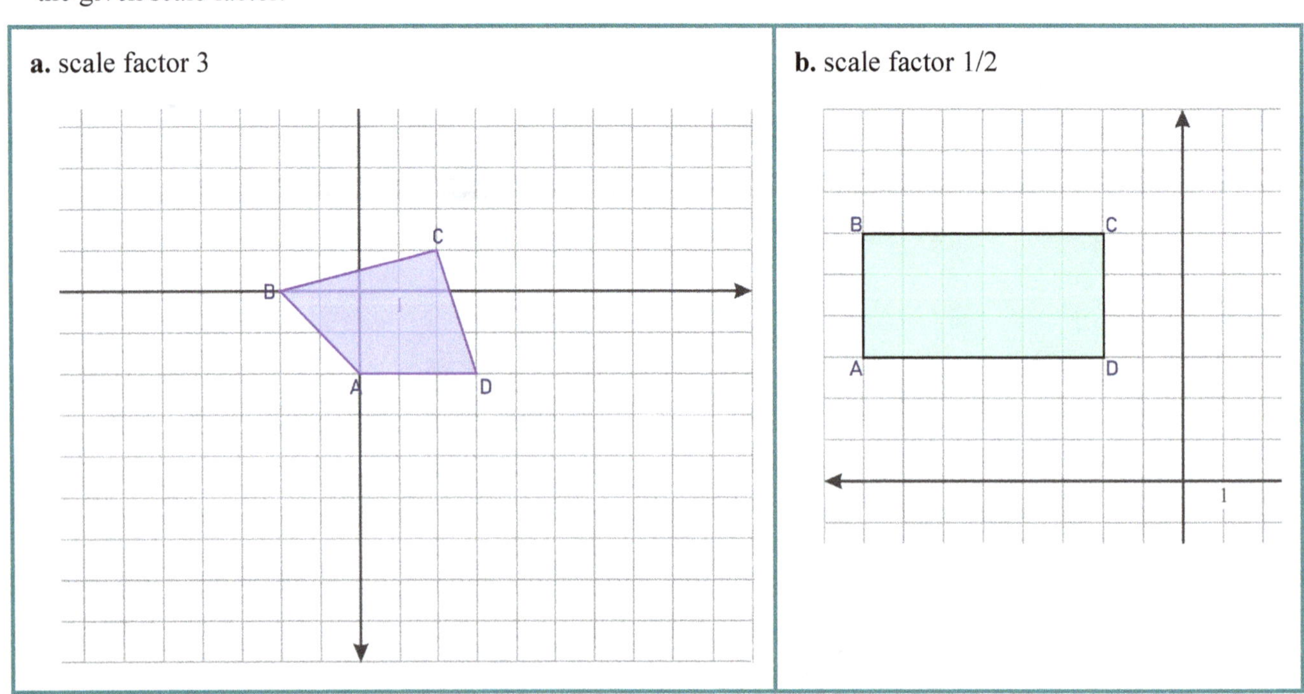

Mixed Review 4

1. Show that the two pentagons are similar by describing a sequence of transformations that maps the larger pentagon to the smaller one.

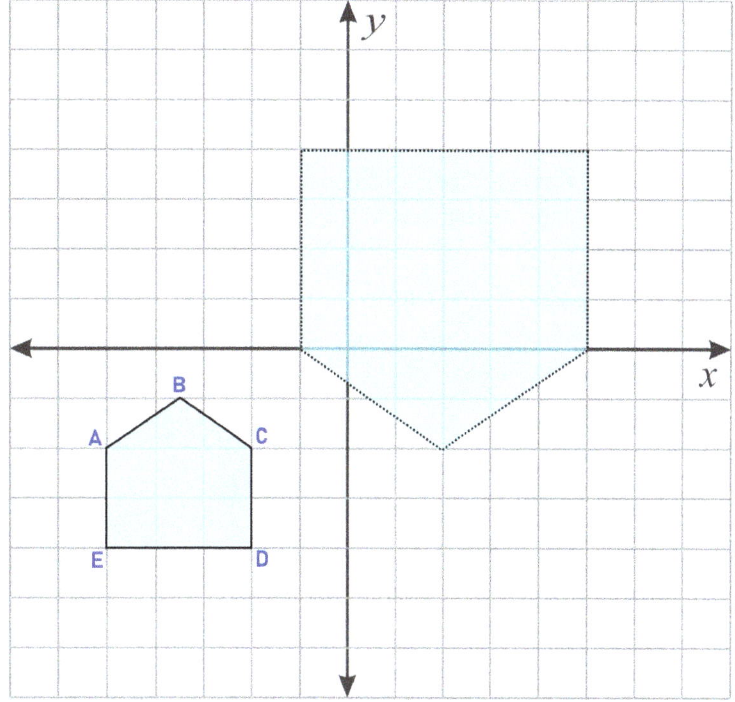

2. Draw a dilation of this triangle from point C and with a scale factor of 1/2.

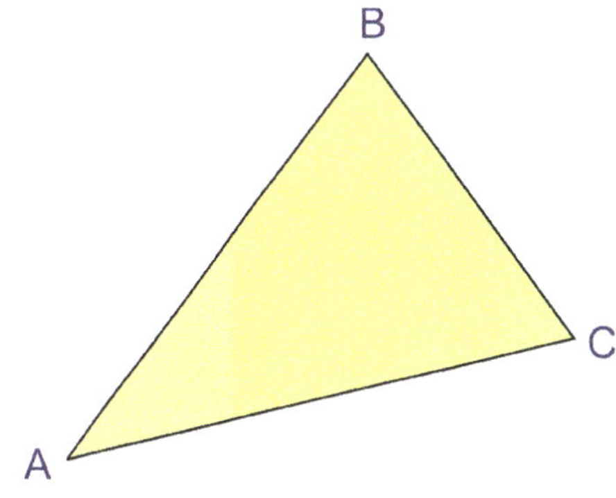

3. Lines *m* and *n* are parallel. Find the measure of the angle marked with "?". Explain your reasoning.

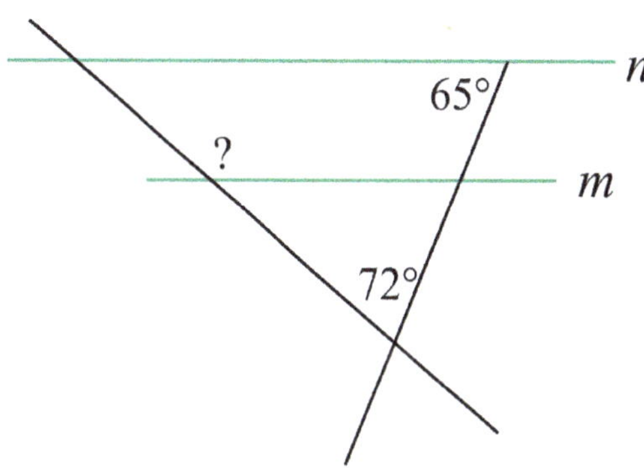

4. Parallelogram ABCD undergoes two congruent transformations, as shown in the image on the right.

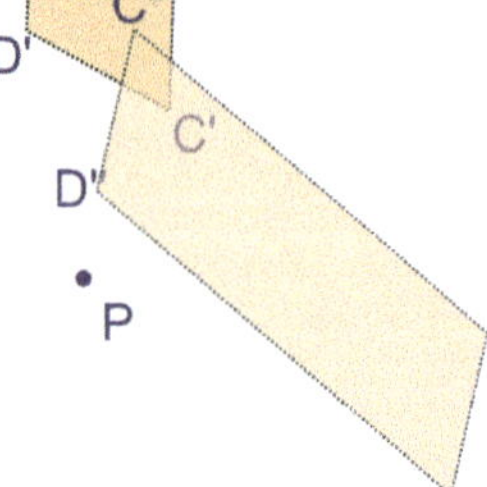

a. Name the two transformations.

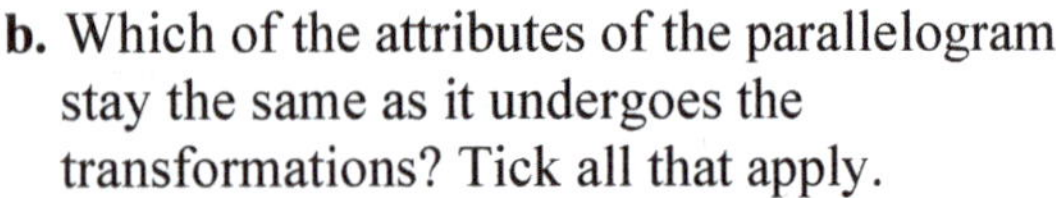

b. Which of the attributes of the parallelogram stay the same as it undergoes the transformations? Tick all that apply.

(i) Perimeter

(ii) Area

(iii) Position

(iv) Angle ABC

(v) Angle sum

5. Solve. Give your answer in scientific notation.

a. $2 \cdot 10^8 + 8 \cdot 10^7$	**b.** $1.3 \cdot 10^6 + 2 \cdot 10^7$	**c.** $3 \cdot 10^5 - 9 \cdot 10^4$	**d.** $2.6 \cdot 10^8 - 5 \cdot 10^7$

6. One of these simplification processes has an error. Which one? Correct it.

a. $\left(\dfrac{4a}{-b^2}\right)^2 = \dfrac{(-4a)^2}{(-b^2)^2} = \dfrac{16a^2}{b^4}$.	**b.** $(2 \cdot 10)^{-3} = \dfrac{2}{10^3} = \dfrac{2}{1000}$

7. A dump truck just dumped 11.0 cubic yards of sand on the ground, and it is in a form of a circular cone now. The diameter of the bottom circle is 12.0 feet. How tall is the cone of sand? 

Note: since one yard is 3 feet, one cubic yard is $3 \text{ ft} \cdot 3 \text{ ft} \cdot 3 \text{ ft} = 27 \text{ ft}^3$.

8. Calculate, and give your answer to a reasonable accuracy.

a. 12.5 km + 2.2 km + 0.054 km = _________ km **b.** 4.6 mi + 409 mi + 34 mi = __________ mi

9. The distance from a campsite to a river is 4.2 cm on a map with a scale of 1:4,000.
How long is this distance in reality, in meters? (Take note of the significant digits in your answer.)

10. How many times does a person's heart beat, in an average lifetime of 70 years? Assume that the average heart rate is 60 beats per minute. Give your answer in scientific notation, and to two significant digits.

11. The Proxima Centauri star is $4.0208 \cdot 10^{13}$ km away from Earth. The speed of light is $3.00 \cdot 10^{5}$ km/s.
How long does the light from this star take to reach Earth? Give your answer in years.

Introduction to Functions Review

1. Change some thing(s) in this table so it represents a function.

Input	Output
Name	**Age**
Fifi	2
Bella	5
Max	
Luna	2
Charlie	6
Luna	3

2. Why is the relationship depicted by the graph *not* a function?

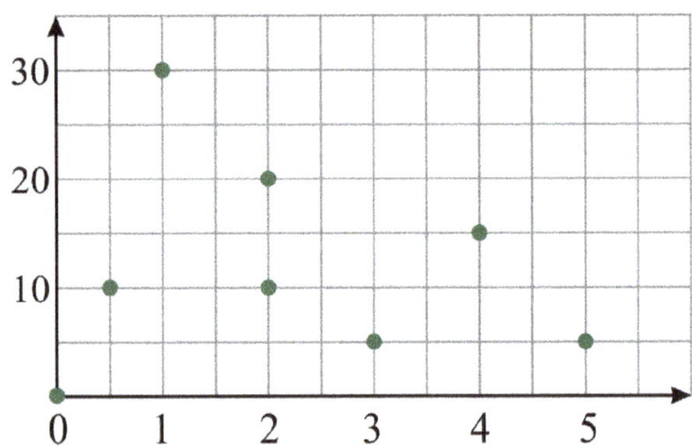

3. Describe this function by intervals where it is increasing, decreasing, or constant. Include also whether it is linear or nonlinear in those intervals.

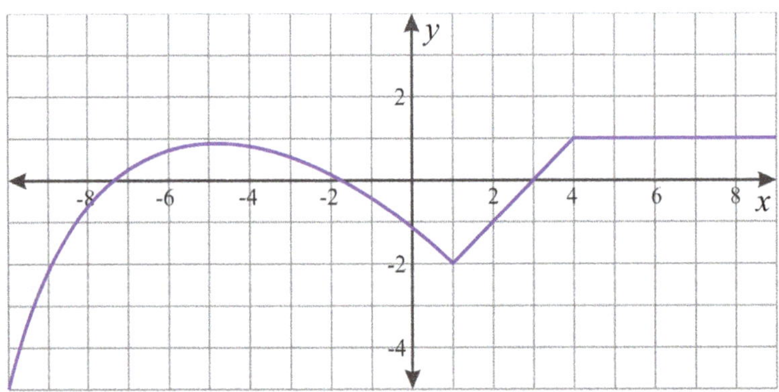

4. Give an example where the cost of something is a function of time, and where the function is *not* linear. Use a table of values to give the function.

Time (hours)	0	1	2	3	4	5	6	7	8	9	10	11	12
Cost ($)													

5. The graph shows the value of a car over time.

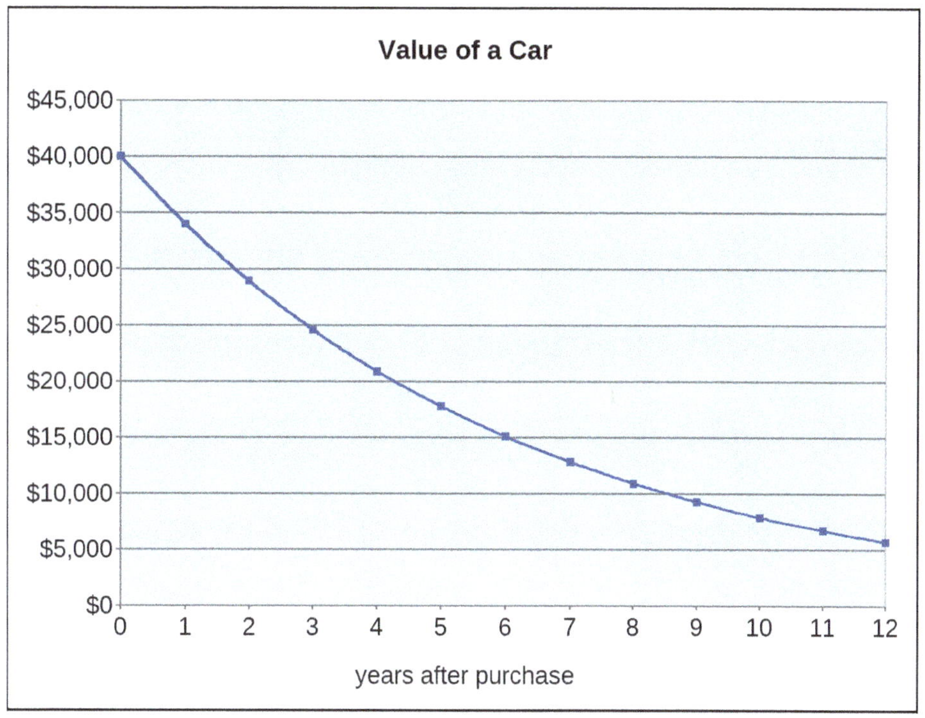

a. Is this a linear function?
How can you tell?

Use approximate values that you can read from the graph, and find the rate of change:

b. from 1 to 2 years after purchase

c. from 6 to 7 years after purchase

6. Jayden and his sister race on bicycles from school back home, a route that is 1.5 miles long. The equation $d = 0.22t$ represents the distance (d, in miles) that Jayden has ridden, as a function of time (t, in minutes). The table below shows the distance his sister has bicycled, at various points in time.

Function 1 — Jayden:

$d = 0.22t$

Function 2 — his sister:

time (minutes)	0	1	2	3	4	5	6
Distance (mi)	0	0.2	0.45	0.68	0.9	1.12	1.24

a. Which function has a greater rate of change from $t = 2$ to $t = 4$ minutes?
What does that represent in terms of real life?

b. Classify each function as either linear or nonlinear.

c. Assume his sister continues with the same speed till the end as what she is riding between 5 and 6 minutes. Who will reach home first?

7. The table below shows the depth of the snow as a function of time from the beginning of a blizzard.

time (hours)	0	1	2	3	4	5	6
depth (inches)	17	20	23	26	29	32	35

a. What is the rate of change?

What does it mean in this situation?

b. What is the initial value?

What does it mean in this situation?

c. Write an equation to represent the relationship between the amount of snow and time in hours.

d. If the blizzard started at 2:30 PM, at what time was the snow 26.5 inches deep?

e. How deep will the snow be at 9 PM?

f. Plot the equation you wrote in (c).

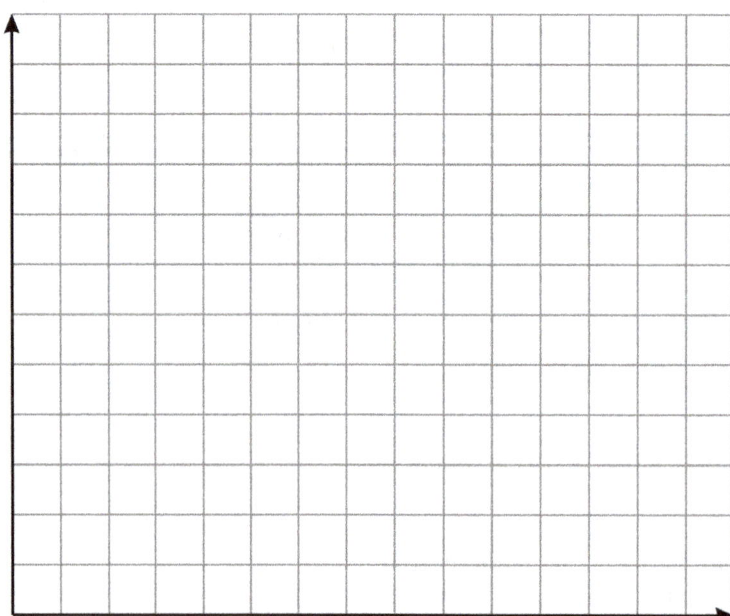

8. Marcia is in math class (in a classroom). The graph shows the distance between her and the blackboard, as a function of time. Make up a story that matches the graph.

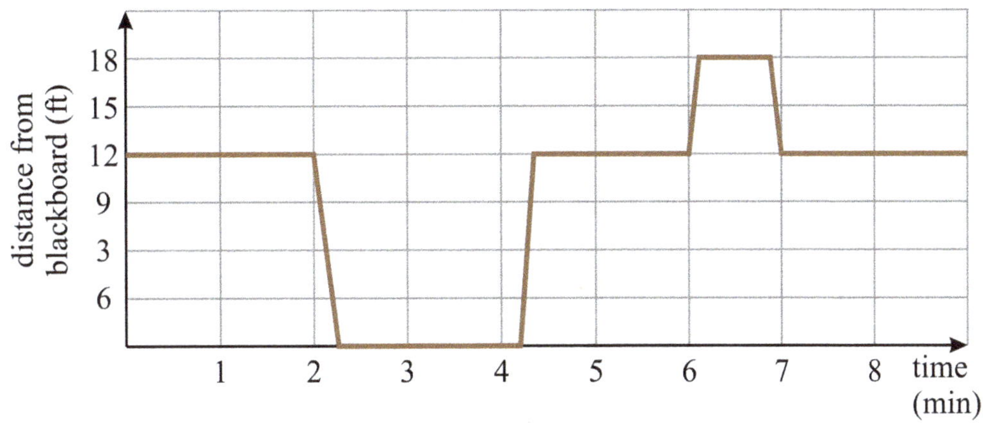

9. Greg is running along his usual running route, and his son Trevor goes with him riding a bicycle, but they don't exactly stay together. The two graphs show the total distance each has covered, as a function of time.

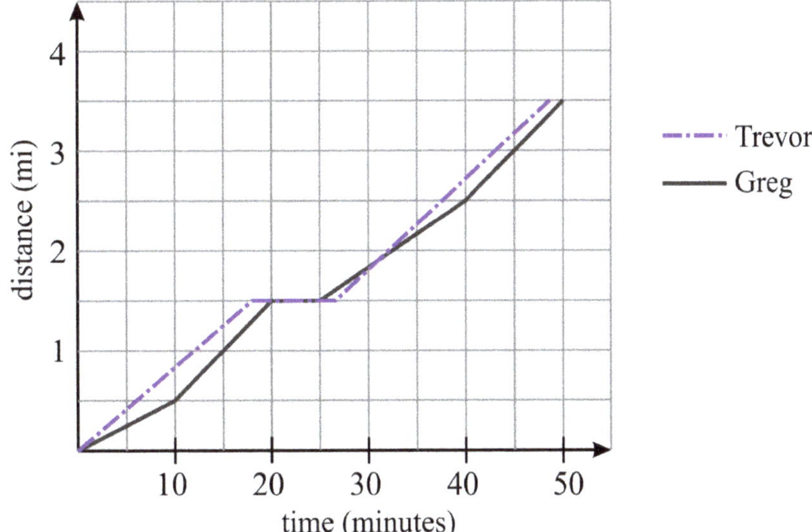

a. From 10-20 minutes, who is going faster?

b. Find two points in time where the two meet. What distance have they traveled at those times?

c. Who finishes the route first? About how much quicker than the other (estimate from the graph)?

Introduction to Functions Test

1. **a.** Change some thing(s) in this table so it represents a function.

   b. If you reversed the inputs and the outputs, would the resulting relationship still be a function? Explain.

Input	Output
Name	***work hours***
Sally	5
Susan	6
Jane	10
Joe	?
Jane	8
Harry	0
Mike	7
Rob	9
John	5

2. Three functions are represented below.

   a. Which one has the largest initial value? Also, state that value.

   b. Which one(s) are linear functions?

   c. Find the rate of change for each function in the x-interval [7, 10].

   Function 1:   Function 2:   Function 3:

Function 1:

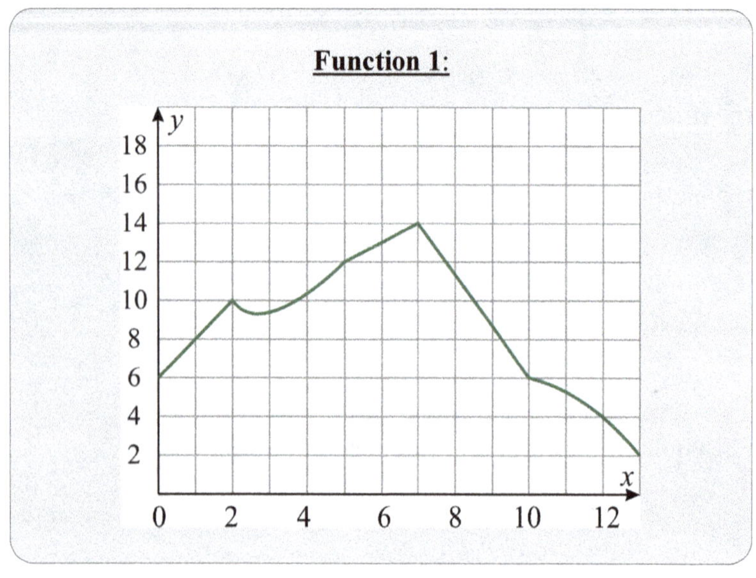

Function 2:

$y = 15 - 1.2x$

Function 3:

x	y
0	16
2	14
4	12
6	11
7	10
8	8
10	5
12	2

3. Describe this function by intervals where it is increasing, decreasing, or constant. Include also whether it is linear or nonlinear in those intervals.

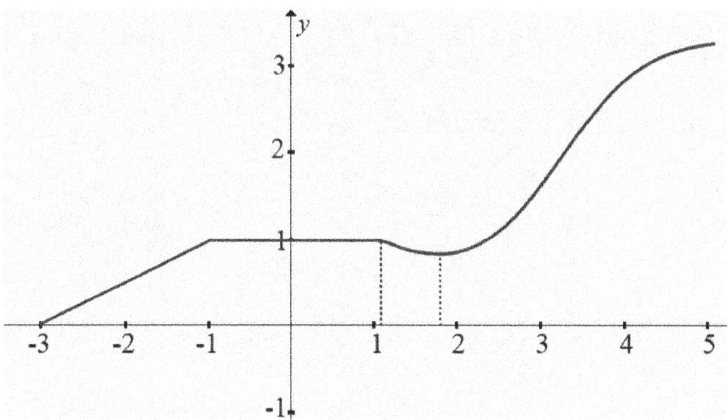

4. Leo is saving money at a steady rate. The equation M = 150*w* + 200 describes the number of dollars he has, as a function of time, where *w* is the number of weeks since January 1.

a. Is this a linear function?

b. What is the rate of change? Include the units.

c. What is the initial value?

d. After how many whole weeks will Leo have at least $1,200?

5. **a.** Sketch a plot of a function that depicts the distance Marsha has walked over time, matching the story below.

For the first 10 minutes, Marsha walks at a fairly brisk and steady pace of 6 km/h. Then she stops for 10 minutes to talk with a friend at a park. Next, she slowly walks a distance of 1 km over the next 15 minutes. She stops again for five minutes. Lastly, she picks up speed and walks at a pace of 8 km/h for 10 minutes.

b. What is the total distance Marsha walked?

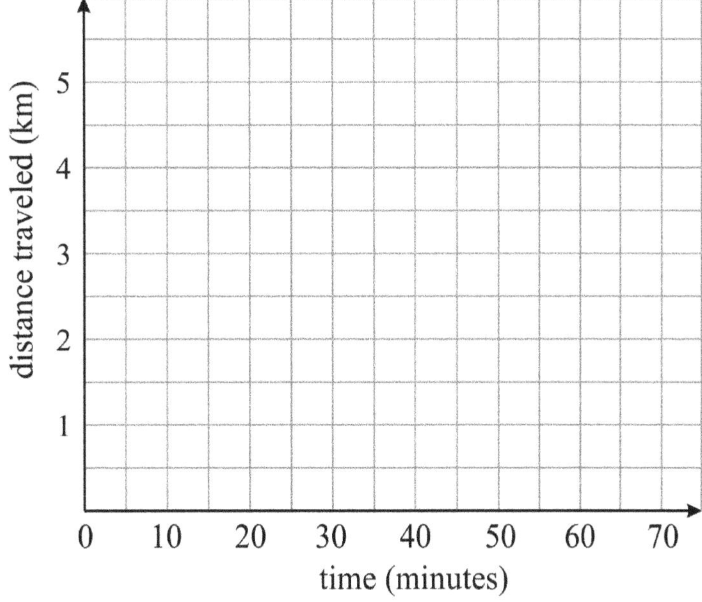

6. Janet starts filling a tub of water with a hose. The volume of water in the tub is a function of time *(t)*.

t **(minutes)**	0	1	2	3	4
V (gallons)	12	16	20	24	28

a. What is the rate of change?

What does it mean in this situation?

b. What is the initial value?

What does it mean in this situation?

c. Write an equation that gives you the volume of water as a function of time.

d. Plot the equation you wrote in (c).

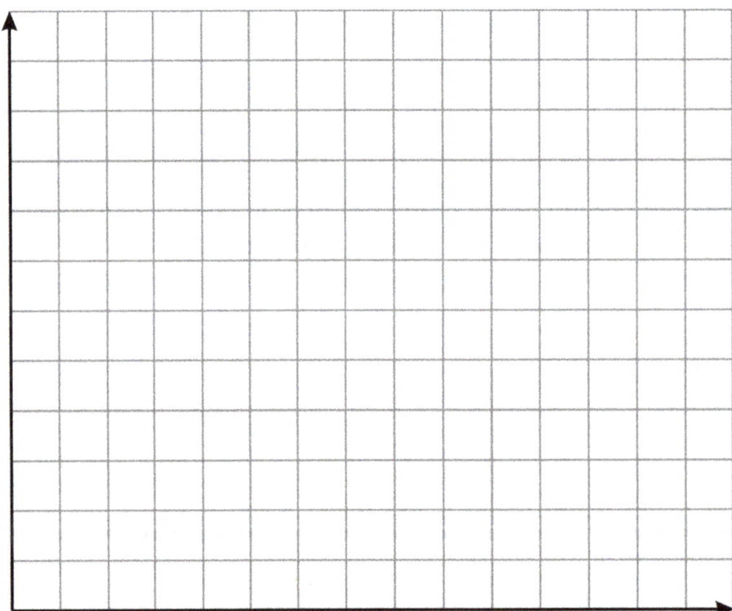

e. How much water is in the tub after 10 minutes?

f. When will the tub have 32 gallons of water in it?

Mixed Review 5

1. Find the expressions that have the value 3^8.

a. $\dfrac{3^{16}}{3^2}$	**b.** 6^4	**c.** 24	**d.** $3^5 \cdot 3^3$	**e.** $\dfrac{3^{11}}{3^3}$	**f.** $\dfrac{3^7}{1/3}$

2. Write an equivalent expression using the exponent laws, without negative exponents.

a. $(b^2)^{-4}$ =	**b.** $(-2y)^4$ =	**c.** $(7a)^{-2}$	**d.** $3x^2x^7y \cdot (-2)y^5$ =
e. $\dfrac{16x^9}{24x^3}$ =	**f.** $\dfrac{5s^2}{s^{-3}}$ =	**g.** $\left(\dfrac{4x}{-6}\right)^2$ =	**h.** $\left(\dfrac{a}{2b^2}\right)^4$ =

3. Solve. Give your answer as a normal number and also in scientific notation.

a. $5 \cdot 10^{-4} + 2 \cdot 10^{-3}$	**b.** $9 \cdot 10^6 + 2 \cdot 10^7$
c. $5 \cdot 10^{-2} - 8 \cdot 10^{-3}$	**d.** $8 \cdot 10^6 - 7 \cdot 10^5$

4. Let $a = 8 \cdot 10^9$ and $b = 2 \cdot 10^7$. Find the value of (a) their product; (b) their quotient. Give your answers in scientific notation.

5. **a.** The volume of a cone is $V = \dfrac{A_b h}{3}$, where A_b is the area of the base and h is the height of the cone. Solve this for h.

b. What is the height of a circular cone with a volume of 20,900 cm^3 and a bottom radius of 25.0 cm?

6. Draw dilations.

a. Draw a dilation of kite DEFG from point E, with scale factor 1/2.

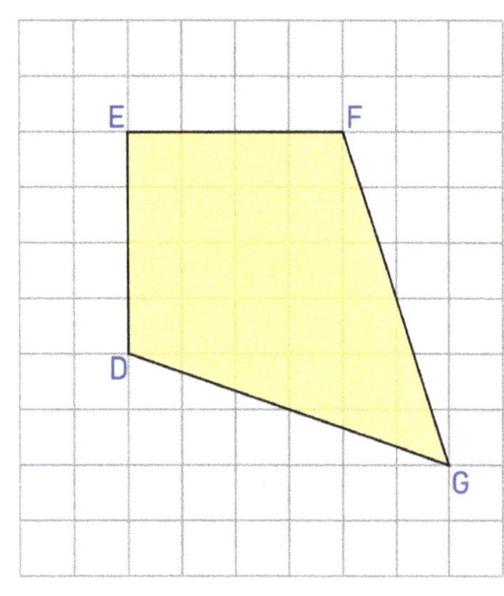

b. Draw a dilation of triangle ABC from point C, with scale factor 3.

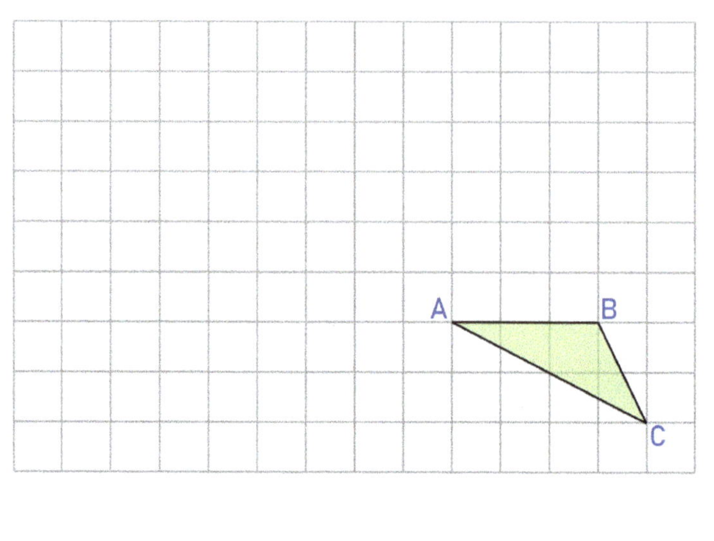

7. The price of gasoline went from \$2.76 to \$2.98 per gallon. What was the percent of increase? 

8. An online curriculum that normally sells for \$72.95 is on sale for 30% off. A sales tax is added to your total, and you pay \$53.92. What is the sales tax percentage? 

9. Solve. Note carefully how the two equations (a) and (b) differ, and how that affects the solution process.

a. $3x + \dfrac{x+3}{5} = 1$	**b.** $3x - \dfrac{x+3}{5} = 1$

10. Tell, without fully solving the equations, whether each equation has one unique solution, no solution, or an infinite number of solutions.

a. $y = 6 - 7y$

b. $6 - 7y = 2 - 7y$

c. $-7y + 14 = 7(2 - y)$

d. $-7y - 2 = -2$

11. Solve.

a. $10 + 3(a + 5) = 2(a - 6) - 4a$	**b.** $20x - 2(x + 1) = 10 - (x - 5)$
c. $\frac{1}{6}x - 1 = 1 + \frac{4}{5}x$	**d.** $2z + \frac{2}{5} = \frac{1}{4}z - 1$

Mixed Review 6

1. Figure DEFG underwent a dilation, then a reflection. Study the coordinates to find out the details about each transformation, then fill in the missing coordinates.

Original figure	Dilation	Reflection
D(−5, −4)	D'(___ , ___)	D"(−2.5, 2)
E(−6, −2)	E'(−3, −1)	E"(___ , ___)
F(−1, −2)	F'(___ , ___)	F"(−0.5, 1)
G(−2, −4)	G'(−1, −2)	G"(−1, 2)

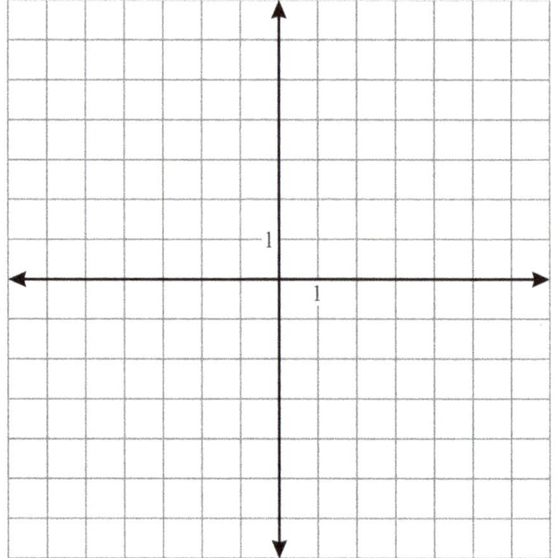

2. A kite was first rotated 180° around the origin, then translated three units to the right, and lastly reflected in the x-axis. Now its vertices are at points (6, 4), (6, 6), (4, 6), and (2, 2).

   a. What were the coordinates of its vertices before these transformations?

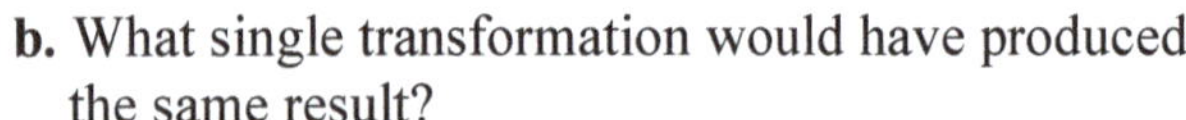

   b. What single transformation would have produced the same result?

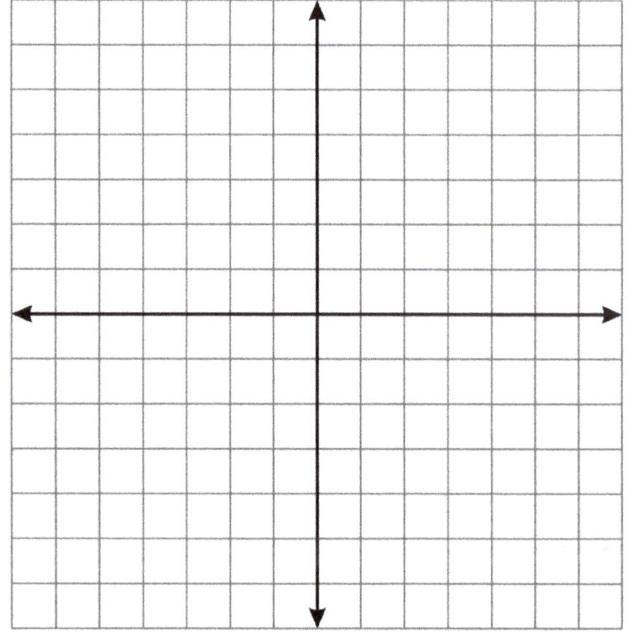

3. An item had three price increases: for 4%, for 7.5%, and for 5%. Now it costs $136.45. What was its price before these increases?

4. Lines m and n intersect at point P. Are the triangles ABP and PQR similar? Explain your reasoning.

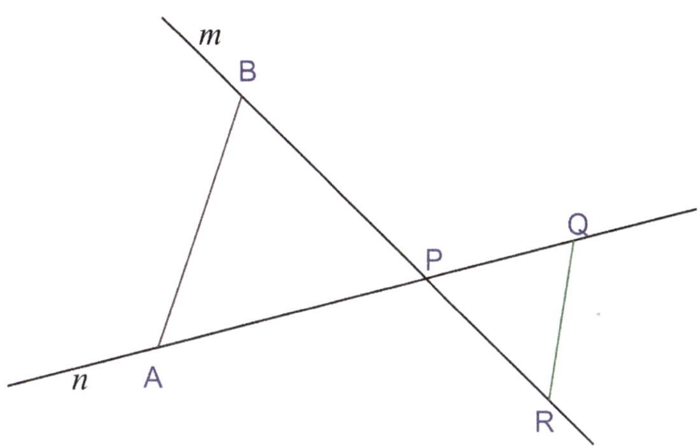

5. Line m is parallel to the line segment $\overline{AB}$. Find the measures of angles x and y.

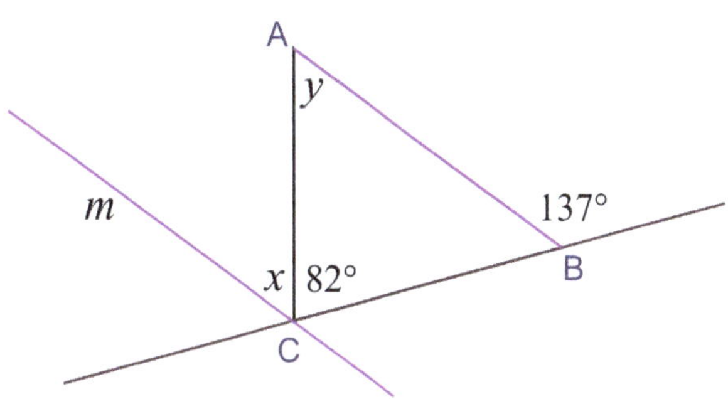

6. You and your friend are in a hardware store, and you are considering buying a low-priced 4-gallon bucket, in the shape of a circular cylinder. Your friend says, "I doubt four gallons of water can fit in that thing; they're cheating you."

You have a tape measure so you quickly measure the diameter of the bottom (it is 11.5 in) and the height (it is 11 in). Should you believe your friend? Explain your reasoning.

Note: 1 gallon = 231 cubic inches

7. Convert. Remember to round your answer to the same amount of significant digits as the measurement.

a. 230 cm = _________ in **b.** 54 L = ___________ gal

c. 24.5 ft = ____________ m **d.** 437 in = __________ m

1 inch = 2.54 cm

1 ft = 0.3048 m

1 gal = 3.785 L

8. One molecule of water consists of two hydrogen atoms and one oxygen atom (H_2O). The mass of one water molecule is about $2.99 \cdot 10^{-23}$ grams. How many water molecules are in 100 grams of water?

9. What errors are made in these solutions? Correct them, and continue the solutions.

a. $5 + x + \dfrac{x-3}{2} = 6 \quad \Big| \cdot 2$

$10 + x - 3 = 12$

b. $x - 2 = 2x - \dfrac{x-1}{5} \quad \Big| \cdot 5$

$5x - 10 = 10x - x - 1$

10. The formula $m = \dfrac{a_1 + a_2 + a_3 + a_4}{4}$ gives you the mean (average) of four numbers a_1, a_2, a_3, and a_4.

a. Solve this formula for a_2.

b. Robert has gotten 67, 85, and 76 points on three math tests, and he has one more to go. He wants his average for the three tests to be 80 (at least). What should he get on his fourth test in order to achieve that?

11. The sum of three consecutive even whole numbers is 13,788. What are the numbers?

Graphing Linear Equations Review

1. Refrigerator companies make estimates of how much energy their fridges consume in typical usage. The table shows how many kilowatt-hours (kWh) of energy fridge 1 consumed over time, and the graph shows the same for fridge 2.

Fridge 1

time (mo)	energy (kWh)
2	75
4	150
6	225
8	300
10	375
12	450

Fridge 2

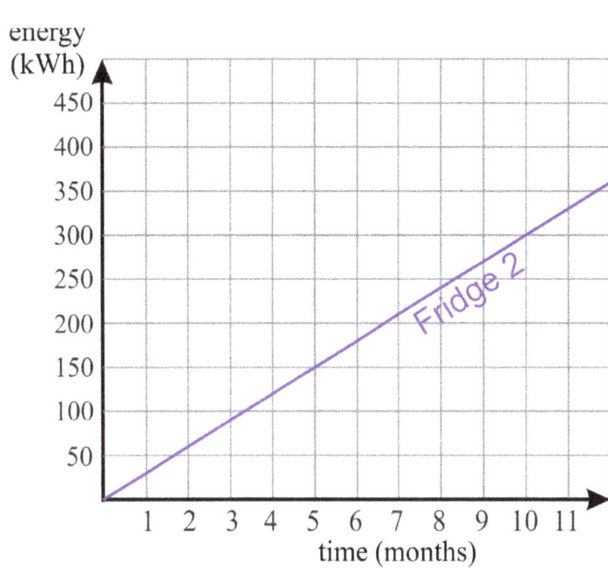

a. Which fridge consumes more electricity in a month?

How much more?

b. Write an equation for each fridge, relating the energy (E, in kWh) and the time (t, in months).

c. Plot the equation for Fridge 1 in the grid.

d. Plot the point corresponding to the unit rate, for Fridge 1.

2. **a.** Find the equations of the four lines, in slope intercept form.

b. (optional) Find the area of the rectangle.

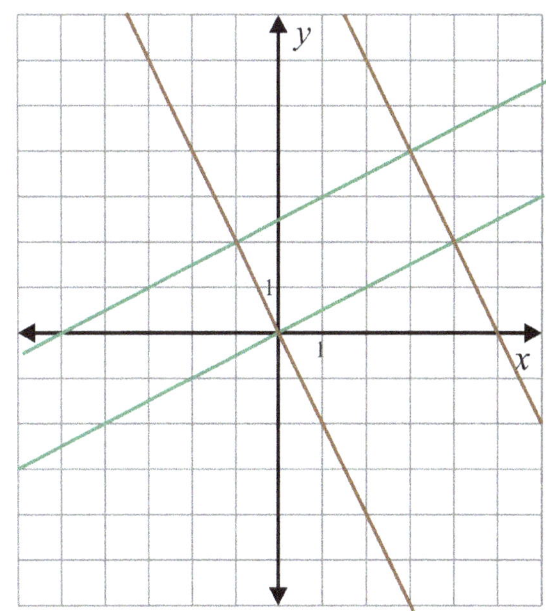

3. Find the equation of each line, in slope-intercept form:

a. has slope 3/4 and passes through $(-2, 3)$

b. is horizontal and passes through $(9, -10)$

4. Find the slope of the lines.
Notice the scaling.

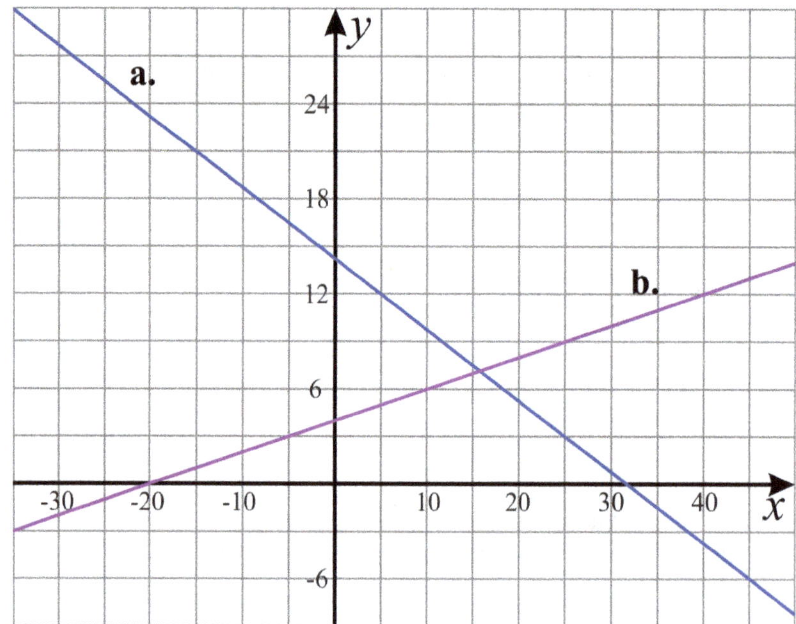

a.

b.

Now find the equations for the lines.

a.

b.

5. Do the three points fall on one line? Explain your reasoning.

$(-3, 1), (-1, -4), (1, -8)$

6. Find s so that the point $(s, 12)$ will fall on the same line as the points $(3, 9)$ and $(15, 18)$.

7. Line S passes through $(-5, -2)$ and $(0, 4)$. Line T is perpendicular to Line S, and passes through $(1, 1)$.

a. Find the equation of line T, in slope-intercept form.

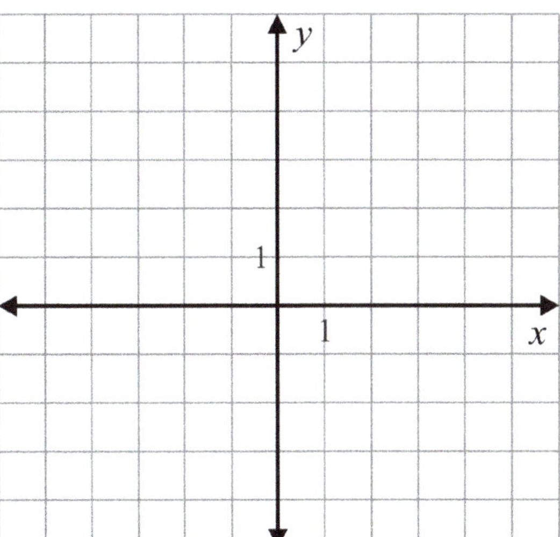

b. Write the equation also in the standard form.

8. Mr. Henson runs a garbage pick-up business, with 12 garbage trucks. To run one truck costs him \$1,500 per month in maintenance costs, plus \$110 a day for fuel.

Consider the cost of running one truck as a function of time, in days (during one month only). Is this a linear relationship, a proportional relationship, or neither?

Write an equation for it.

9. Match the descriptions and the equations.

$y = (-4/3)x - 7$	Is parallel to $x = 9$ and passes through $(2, 7)$
$3x - y = -21$	Has y-intercept -4 and is perpendicular to $y = -2x$.
$y = -4$	Passes through $(-5, 6)$ and has slope 3.
$x - 2y = 8$	Passes through $(-9, 5)$ and $(-3, -3)$
$x = 2$	Passes through $(-3, 0)$ and $(0, 9)$
$y = 3x + 9$	Has y-intercept -4 and is parallel to $y = -2$.

10. Transform each equation of a line to the standard form, and then list its x and y-intercepts.

a. $y - 6 = 2(x + 2)$	**b.** $-\frac{1}{3}x - \frac{3}{2}y = 1$

11. A heater was turned on at 10 AM in a cold, uninhabited house, to prepare it for people later that day. The graph shows the temperature of the house. The count of hours starts at 10 AM. 

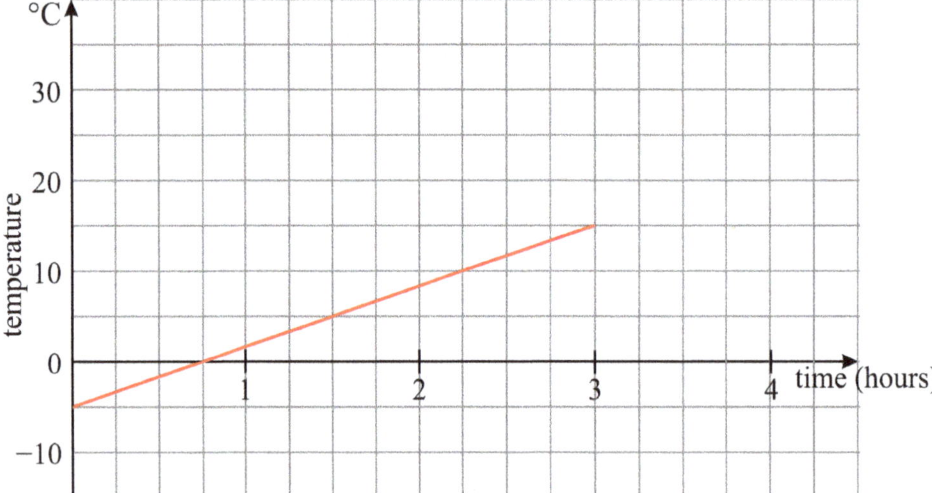

a. Write an equation for the line.

b. If the temperature continues to rise in the same fashion, what will the temperature be at 2:30 PM?

c. When will the temperature reach 22°C?

d. Let's say the heater is turned off at 1:45. What is the temperature at that time?

e. If the house had started out at a temperature of −12°C instead, and the heating process worked in the same fashion (the temperature rose at the same rate), at what time would the house reach a temperature of 22°C?

Graphing Linear Equations Test

1. Liam is buying paint for his house. He is considering two different types of paint, Paint 1 and Paint 2. The equation $A = 12p$ gives you the area (A), in square meters, that p liters of Paint 1 covers. The graph shows how much area Paint 2 covers.

Paint 1: $A = 12p$ **Paint 2:**

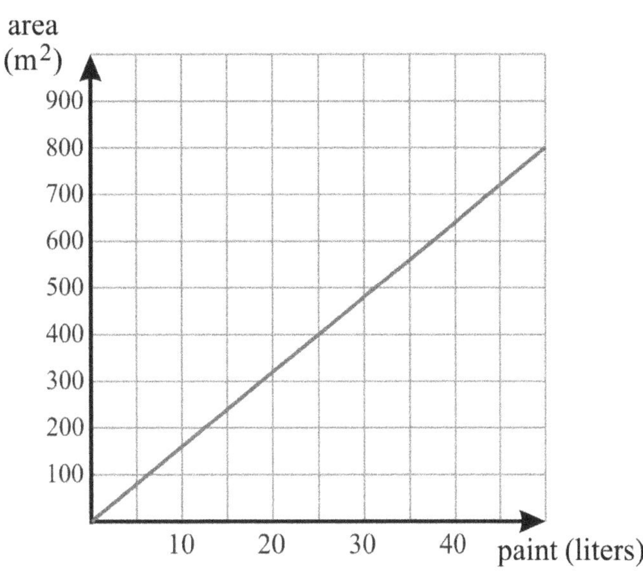

a. What is the slope of the line depicting the coverage of Paint 2?

b. Which paint covers more area per liter?

c. The walls of a bedroom are 30 m^2. Calculate how much of each kind of paint would be needed to paint the room.

d. Draw a line depicting the area that Paint 1 covers, in the same grid as for Paint 2.

e. How much more does 40 liters of the one paint cover than 40 liters of the other?

2. Find the equation of each line, in slope-intercept form:

a. has slope 3/4 and passes through (0, −8)

b. has slope −5/2 and passes through (−2, 3)

c. is vertical and passes through (−11, 4)

d. is parallel to $y = 6x - 9$ and crosses the line $x = 4$ at $y = 2$.

3. Henry has been helping his obese cat to lose weight. He used to weigh 7.6 kg, but started losing weight in a steady manner so that a month later, he weighed 7.2 kg and two months later 6.8 kg.

a. Consider the cat's weight as a function of time, and write an equation for it.

b. Graph your equation. Design the scaling for the vertical axis so that the point (0, 7.6 kg) fits in it.

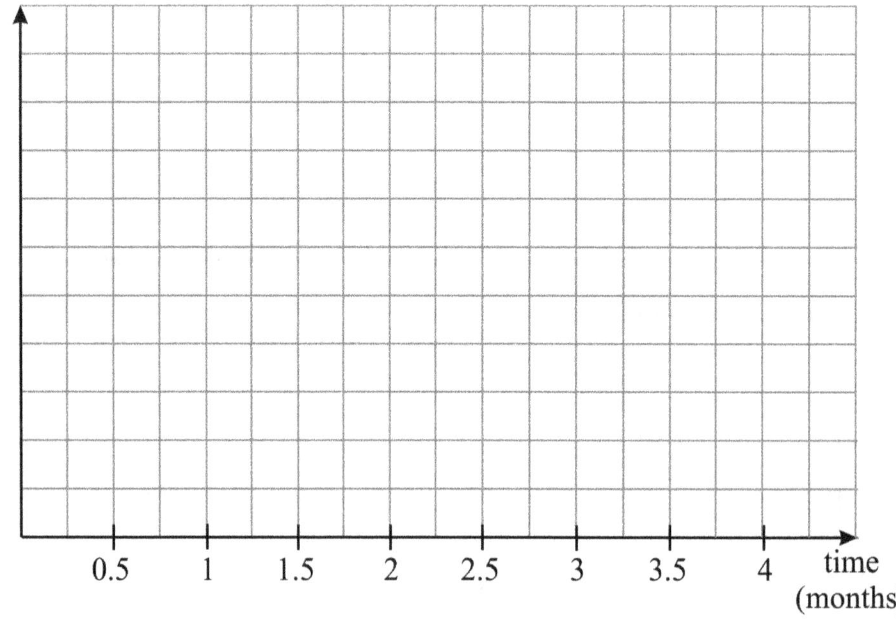

c. In how many months will the cat weigh 5 kg if this continues?

4. Find the equations for the lines. Notice the scaling.

a.

b.

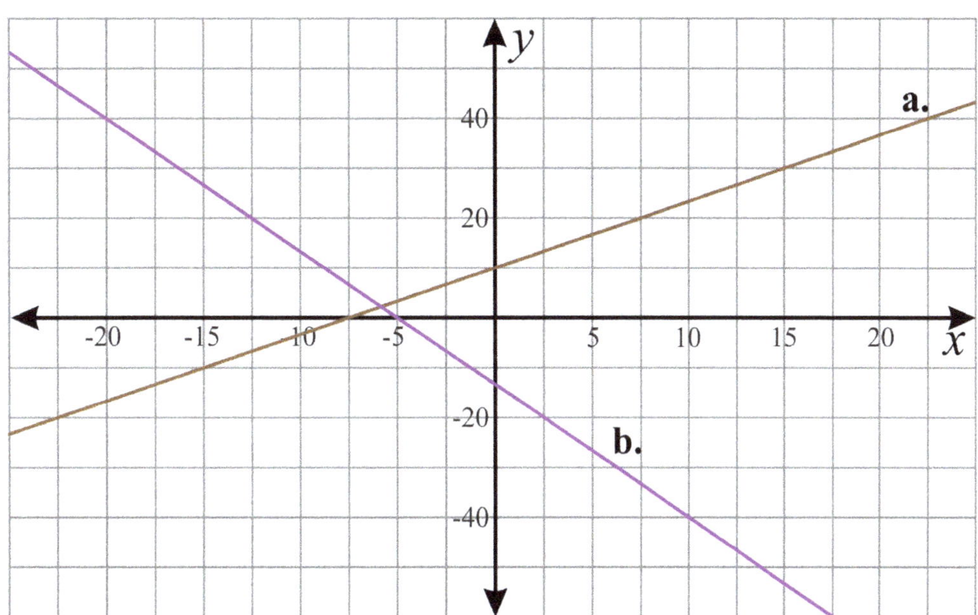

5. Write the equation $9x - 3y = 6$ in the slope-intercept form, and graph it.

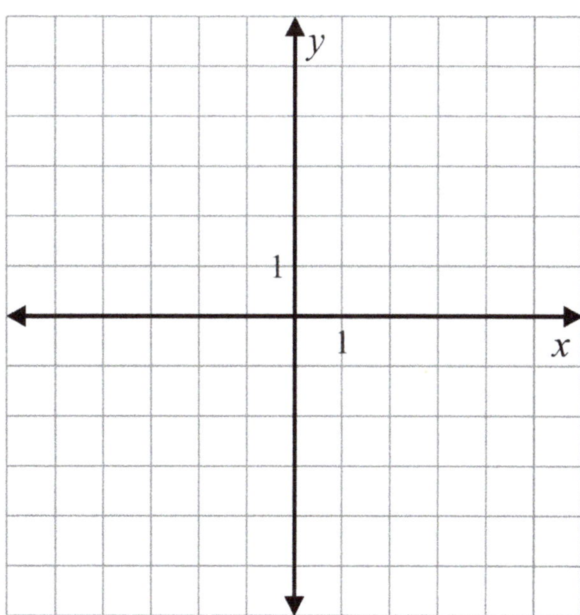

6. Line L passes through (2, −1) and (−3, 6). Line M is perpendicular to Line L, and passes through (−2, −3).

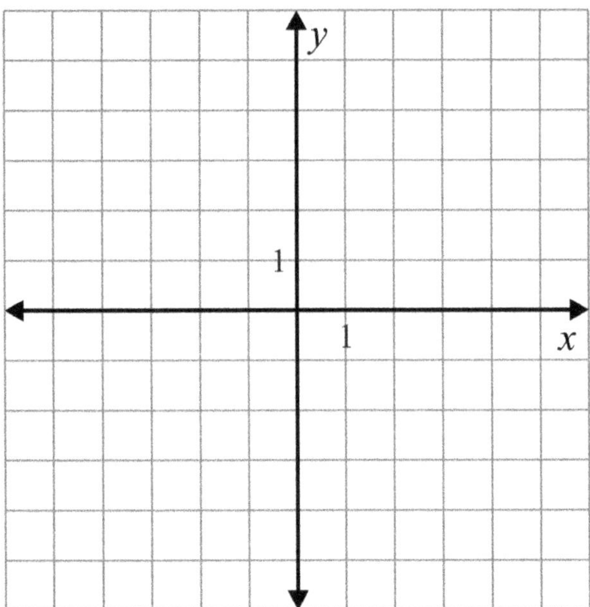

a. Find the equations of both lines, in slope-intercept form.

b. Write the equations also in the standard form.

7. Find s so that the point (6, s) will fall on the same line as the points (1, −1) and (10, −28).

Mixed Review 7

1. The graph shows the distance between Tyler and his home, as he goes to the grocery store and comes back. Continue graphing the function. Here is the story:

   - Tyler leaves home, driving at a constant speed of 60 km/h.
   - After five minutes, he hits a traffic jam and sits there without moving for ten minutes.
   - After that, he starts moving again at a speed of 42 km/h, and continues with that for 10 minutes.
   - Then he reaches the store, and stays there for 30 minutes.
   - On his way home, he is able to keep a steady speed of 48 km/h.

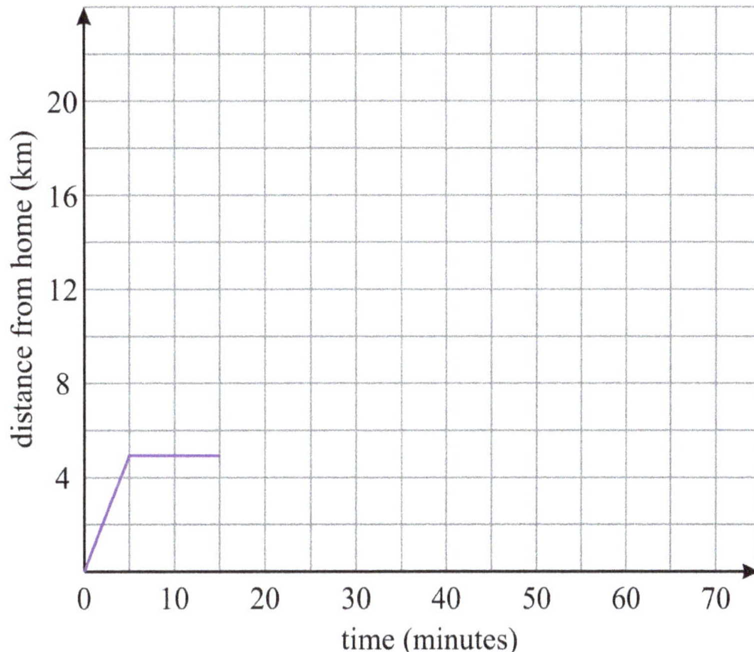

2. Prove that figure 3 is congruent to figure 4 by explaining a sequence of transformations that maps figure 3 onto figure 4.

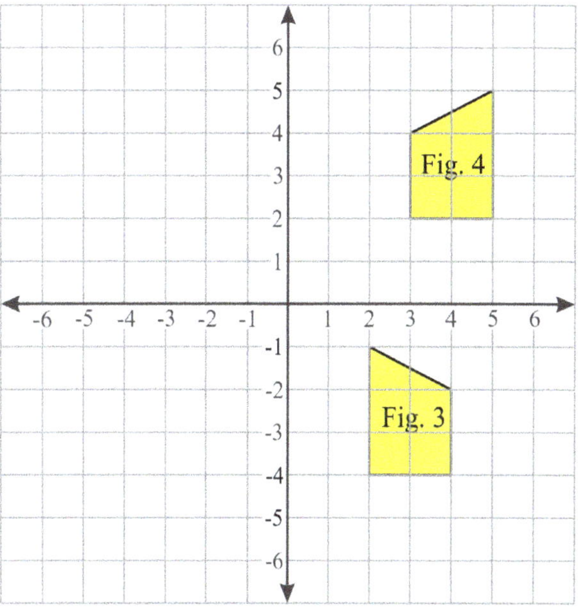

3. **a.** How many solutions does this equation have?

$$3(y - 1) = 7 + 3y$$

b. Modify the equation so that it has *one* solution.

4. Two lines intersect at point C.

a. Are the triangles ABC and CDE similar?
How do you know?

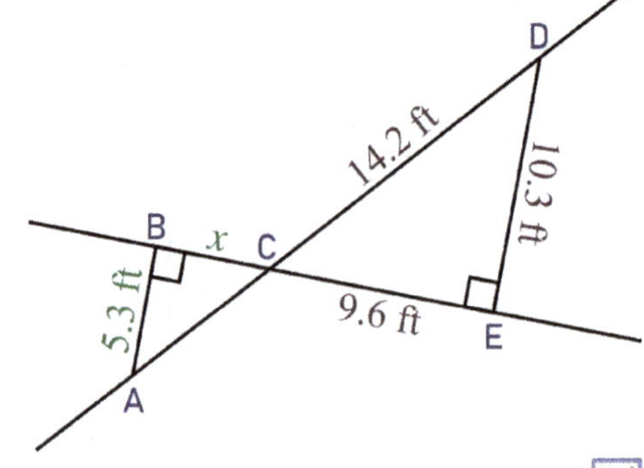

b. Find the value of x.

5. Which of the three functions represented below has the largest rate of change...

a. in the x-interval [1, 3]?

b. in the x-interval [4, 5]?

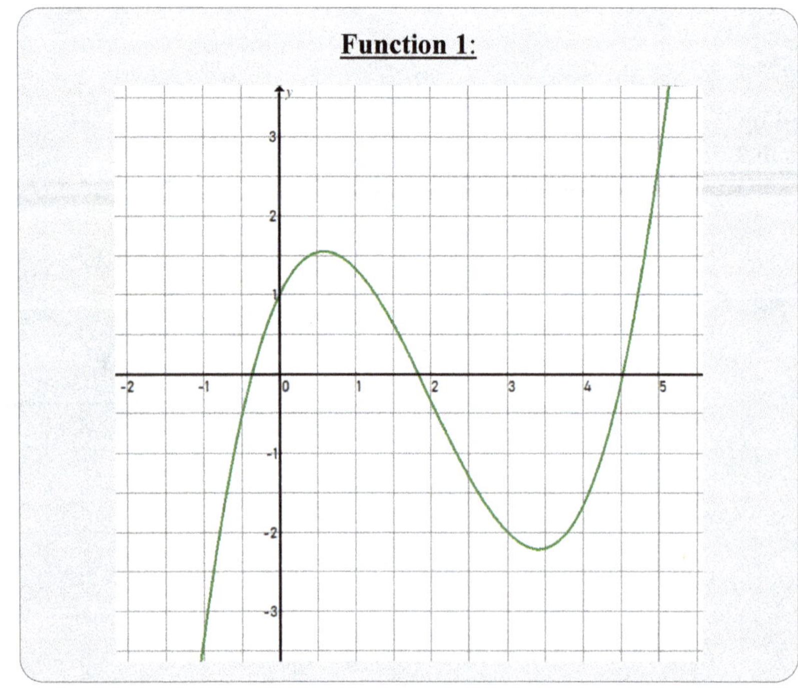

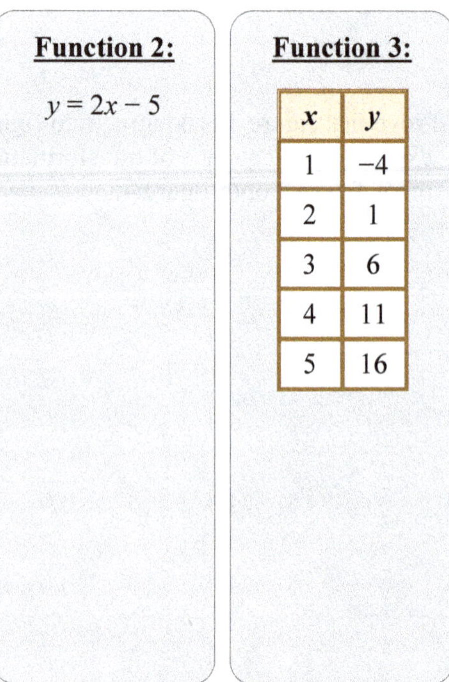

Function 2:

$y = 2x - 5$

Function 3:

x	y
1	−4
2	1
3	6
4	11
5	16

6. Solve.

a. $10(x-3)+2x-5 \;=\; 6-3x$	**b.** $\frac{1}{3}x-5 \;=\; \frac{1}{4}x+2$
c. $\frac{1}{6}(x-7) \;=\; -\frac{7}{8}$	**d.** $\frac{5x-2}{10} - 2 \;=\; 3x$

7. Mary says to Ryan, "Twenty years ago, I was 3/5 of your age, and now I am 7/9 of your age." How old is Mary now?

8. Amy's piggy bank has a bunch of nickels, twice as many dimes, and 16 quarters. The total value of her coins is $12.25. How many of each type of coin does she have?

Mixed Review 8

1. For what value of a would the equation $4(x + 7) = 10 - a(2x - 1)$ have *no* solutions?

2. The area of a kite is $A = \frac{pq}{2}$, where p and q are the diagonals of the kite.

   a. Solve this formula for p.

   b. If the area of a kite is $0.6\ m^2$, and the shorter diagonal is 90 cm, how long is the other diagonal?

3. **a.** Ann made a circular cake (in the form of a circular cylinder). Its bottom diameter is 9.0 inches and its height is 4.0 inches. If it is divided into 12 equal slices, what is the volume of one slice?

   b. If a square cake is baked in an 8" by 8" pan, what should its height be so that it would have the same volume as the circular cake?

4. Solve.

a. $2x = \frac{x - 5}{6} - 2$	**b.** $\frac{x - 1}{4} + \frac{2x + 7}{3} = 0$

5. An airplane travels at a constant speed of 600 mi/h from New York to Los Angeles, a distance of 2,450 miles.

a. Consider the distance (d) the airplane has traveled as a function of time (t). Write an equation relating the two variables.

b. What is the rate of change of this function?

c. Is this function linear? How do you know?

d. Plot your equation. Notice that you need to decide a scale for the d-axis.

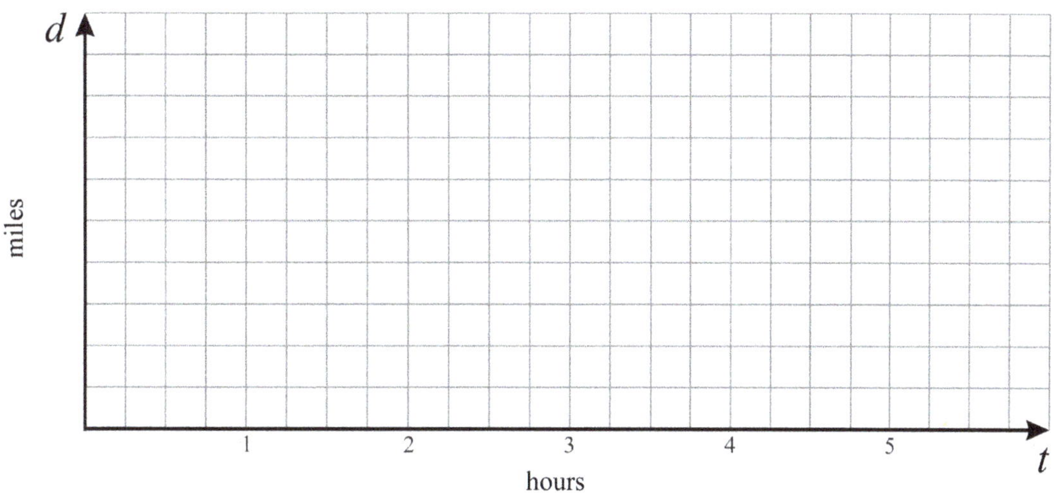

e. How far will the airplane travel in 1 hour 40 minutes?

6. The table shows the distance as a function of time for another airplane.

Time (h)	Distance (mi)
0	0
0.5	300
1	600
1.5	950
2	1,200
2.5	1,500
3	1,750
3.5	2,100
4	2,400

a. Is this function linear?

How do you know?

b. Compare this to the function in question #5. Which function has a greater rate of change from 3 to 4 hours?

7. Simplify, writing an equivalent expression that does not have any negative exponents.

a. $(2x^3)^{-1}$ =	**c.** $(ab^3)^{-2}$ =	**e.** $(-2x)^5 \cdot (5x)^{-1}$ =
b. $(8a^{-5})^2$ =	**d.** $(-3s^{-2}t)^3$ =	**f.** $z^4 \cdot (-3z^3)^{-2}$ =

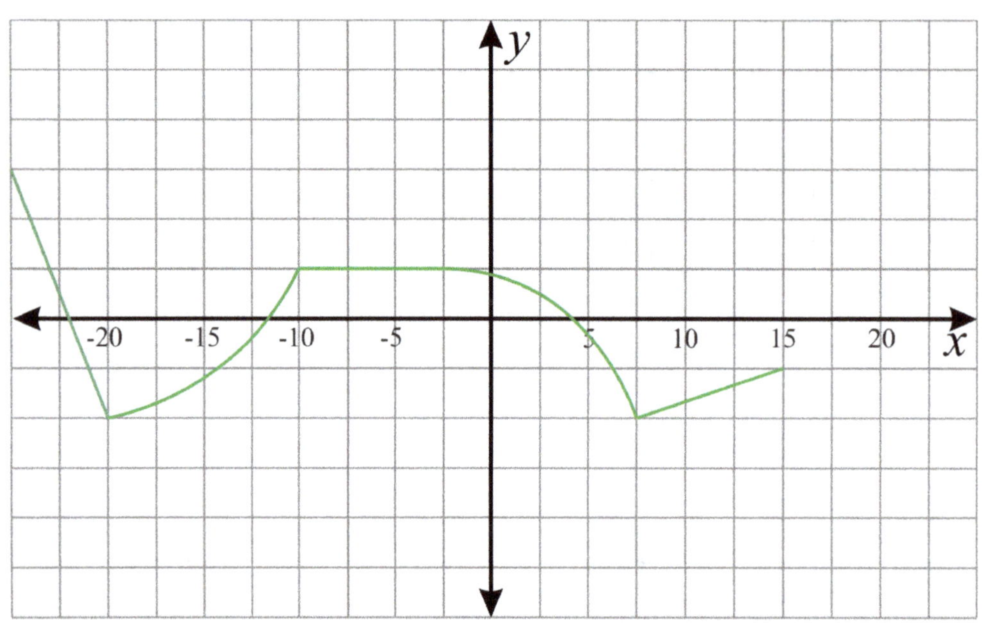

8. Above, you see the graph of a certain function. Describe where this function is

a. linear and decreasing

b. constant

c. nonlinear and decreasing

9. Continue drawing the graph of the function in #8 from $x = 15$ till $x = 25$ so that it is nonlinear and increasing.

Irrational Numbers and the Pythagorean Theorem Review

1. Find the values of these (principal) square roots and cube roots.

a. $\sqrt{64}$	**b.** $\sqrt{169}$	**c.** $\sqrt{2{,}500}$	**d.** $\sqrt{0.81}$
e. $\sqrt{\frac{36}{100}}$	**f.** $-\sqrt{49}$	**g.** $\sqrt[3]{125}$	**h.** $\sqrt[3]{27{,}000}$

2. Between which two whole numbers do the following square roots lie? Do not use a calculator.

a. $\sqrt{7}$ **b.** $\sqrt{77}$ **c.** $\sqrt{134}$ **d.** $\sqrt{48}$

3. Find the value of $\sqrt{13}$ to one decimal digit, without a calculator.

4. Plot the following numbers *approximately* on the number line. Do not use a calculator, but think between which two integers the square root lies, and whether it is close to one of those integers, using mental math.

$\sqrt{50}/2$ $\sqrt{17}$ $-\sqrt{8}$ $\sqrt[3]{8}$ $\sqrt{101} - 4$ $-\sqrt[3]{27}$ π $-\sqrt{38}/3$

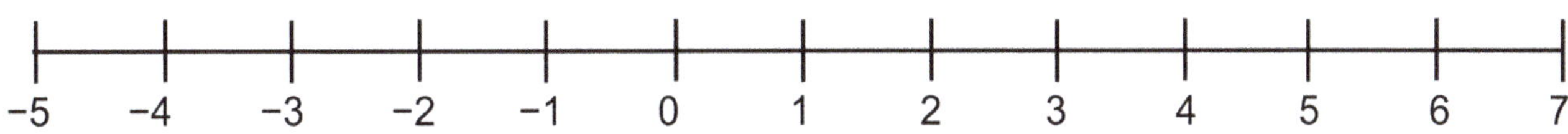

5. Compare, writing >, <, or = between the numbers. Think between which two whole numbers the square root lies, using mental math.

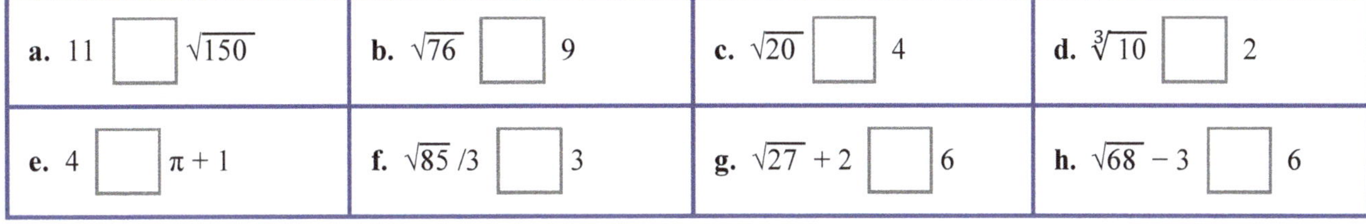

a. $11 \ \square \ \sqrt{150}$	**b.** $\sqrt{76} \ \square \ 9$	**c.** $\sqrt{20} \ \square \ 4$	**d.** $\sqrt[3]{10} \ \square \ 2$
e. $4 \ \square \ \pi + 1$	**f.** $\sqrt{85}/3 \ \square \ 3$	**g.** $\sqrt{27} + 2 \ \square \ 6$	**h.** $\sqrt{68} - 3 \ \square \ 6$

6. Find the value of the expressions.

a. $\sqrt{144}$	**b.** $-\sqrt{81}$	**c.** $\sqrt{1{,}600}$
d. $\sqrt{10^2 - 6^2}$	**e.** $\sqrt{49 \cdot 49}$	**f.** $\sqrt{5 \cdot (83 - 3)}$

7. **a.** If the side of a square measures $\sqrt{7}$, what is its area?

b. What is the perimeter of a square with an area of 20 square units?
Give your answer as an exact value (not rounded).

8. Determine whether the following numbers are rational or irrational, and explain why.

a. 0.8053

b. $\sqrt{2{,}500}$

c. 5π

d. $-\sqrt{56}$

e. $-\frac{2}{7}$

f. $\frac{1}{\sqrt{36}}$

g. $2.1\overline{09}$

h. 0.020202

i. $0.20\overline{8}$

j. $4\sqrt{7}$

9. Write each repeating decimal as a fraction.

a. $0.\overline{61}$	**b.** $4.1\overline{7}$

10. Solve. Give the final answers in exact form.

a. $x^2 = 147$	 **b.** $a^2 = 169$	 **c.** $w^3 = 0.36$
d. $3x^3 = 21$	**e.** $5b^3 = 625$	**f.** $2a^3 = -16$

11. Solve. Give your answer to the nearest thousandth. You may use a calculator.

 **a.** $y^2 + 18 = 35$	 **b.** $0.6h^2 = 4$

12. For each set of lengths, determine whether they form a right triangle.

a. 20, 24, 30

b. 2.6, 1.0, 2.4

13. Solve for the unknown side. Leave your answer in root form if the radicand is not a perfect square.

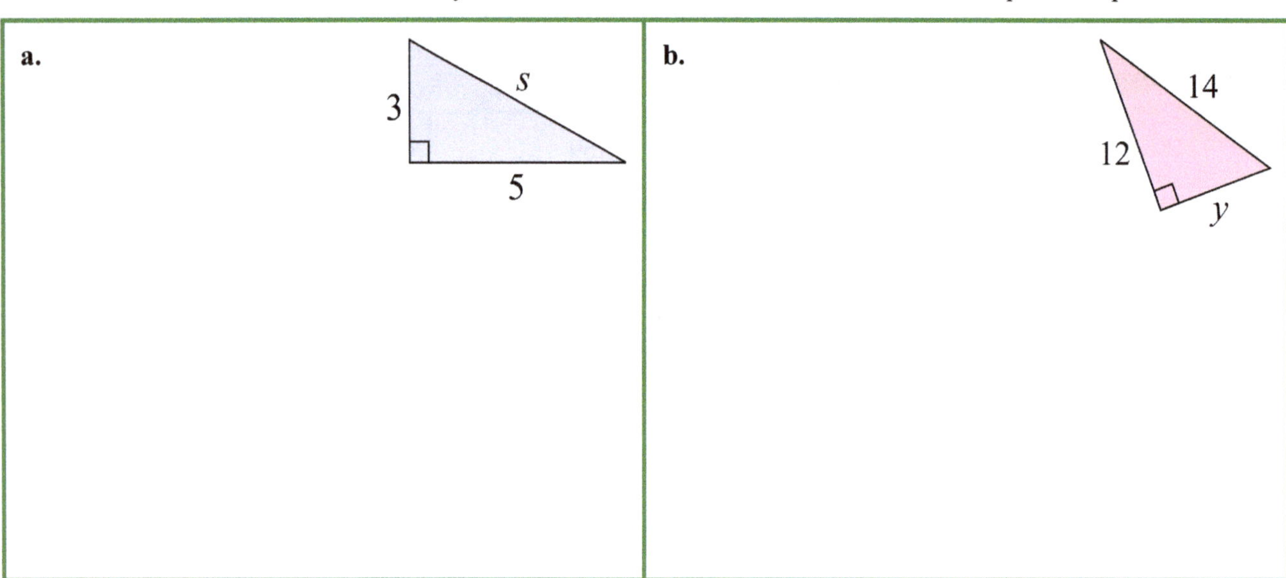

14. Solve for the unknown side. Round your answer to the same accuracy as the given numbers.

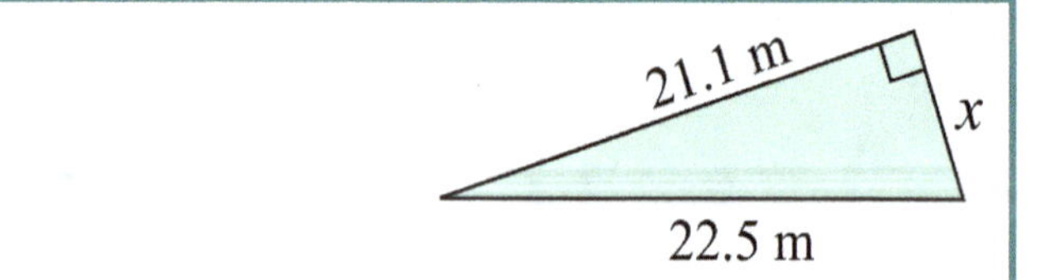

15. The two legs of a right triangle are $\sqrt{7}$ and $\sqrt{8}$.
How long is the hypotenuse?

16. Lauren and Anna want to make this pennant for their jogging club. Calculate its area.

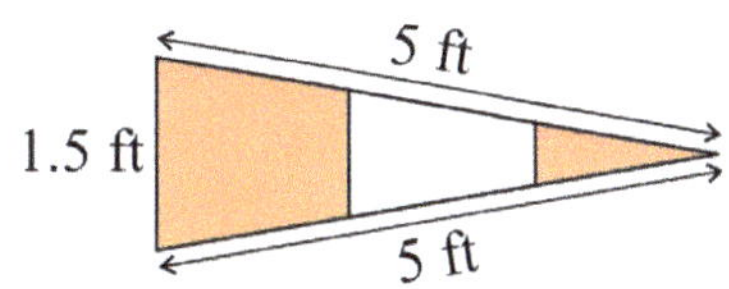

17. Arrange the pieces in the empty square in a manner that will prove the Pythagorean Theorem, and explain how your arrangement does so.

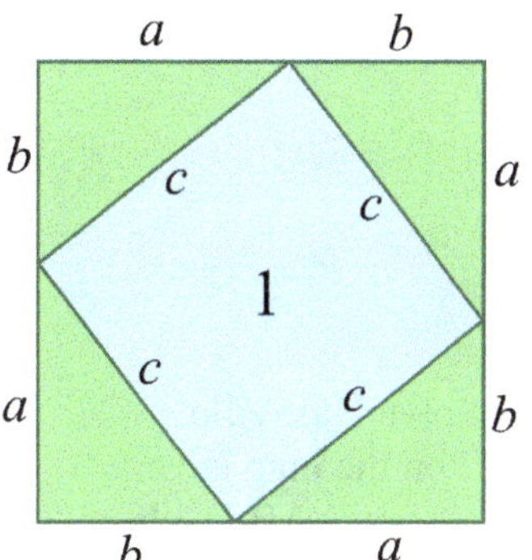

18. Find the area and the perimeter of the garden, if one unit in the grid is 2.0 feet.

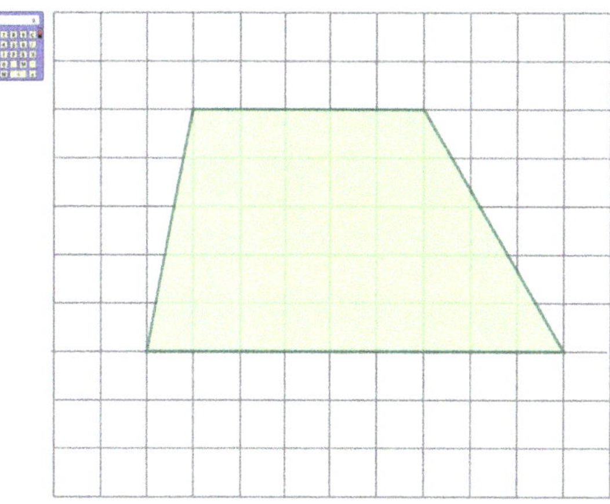

19. Find the distances between the points, to the nearest tenth of a unit.

a. (−3, 4) and (11, −9)	**b.** (42, −15) and (70, 100)

20. Find the diagonal of the cube with edges 2.5 m long.

21. The map shows part of downtown Nashville, Tennessee. The triangle ABC on the map is very close to a right triangle. The distance AB is 370 m and the distance AC is 620 m. However, these distances are approximate, so your calculations will also be only approximate.

About how much shorter is it to travel from point A to point C along Lafayette Street than to travel first along Korean Veterans Boulevard and then along 5th Avenue South?

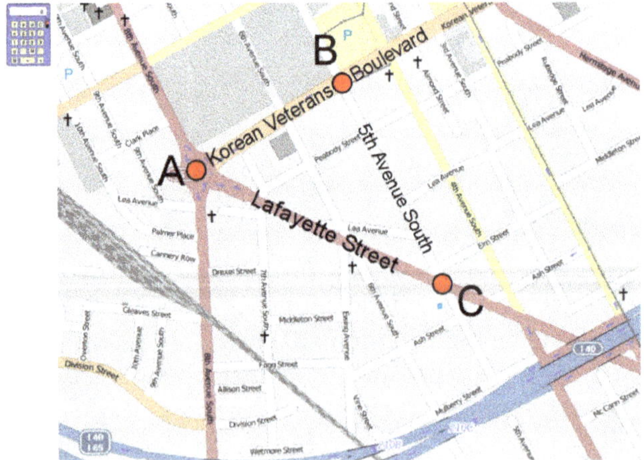

Irrational Numbers and the Pythagorean Theorem Test

A calculator is not allowed for the first page of the test.

1. Find the correct statements.

a. 0.141414 is rational because its decimal expansion repeats.

b. $\sqrt{111}$ is irrational because it is a square root of a number that is not a perfect square.

c. 2π is irrational because π is irrational, and an irrational number multiplied by a rational number is irrational.

d. $\frac{\sqrt{64}}{3}$ is rational because it is a ratio of two whole numbers.

2. Write the two consecutive integers between which the square or cube root lies.

a. _____ $< \sqrt{44} <$ _____	**b.** _____ $< -\sqrt{5} <$ _____	**c.** _____ $< \sqrt[3]{21} <$ _____

3. Plot the following numbers *approximately* on the number line. Do not use a calculator.

$-\sqrt{25}$ $\quad \sqrt{80} - 5$ $\quad \sqrt[3]{27}$ $\quad -2\sqrt{2}$ $\quad \pi$ $\quad \sqrt{35}/2$

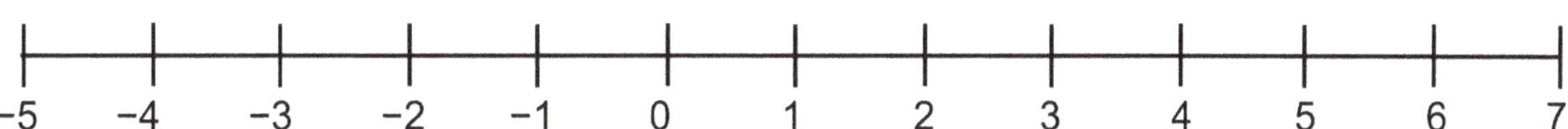

4. Solve. Give the final answers in exact form.

a. $x^2 = 37$	**b.** $5a^2 = 90$	**c.** $w^3 = 125$

5. Write the repeating decimal $0.\overline{23}$ as a fraction.

A calculator is allowed for the remaining portion of the test.
Please return the first page of the test to your teacher before continuing.

6. The two legs of a right triangle are $\sqrt{21}$ and $\sqrt{15}$. How long is the hypotenuse?

7. Find the distance between (−2, 5) and (10, 8). Give your answer rounded to two decimals.

8. Elizabeth runs around this track for morning exercise. Find the total distance she runs if she does three loops around this track.

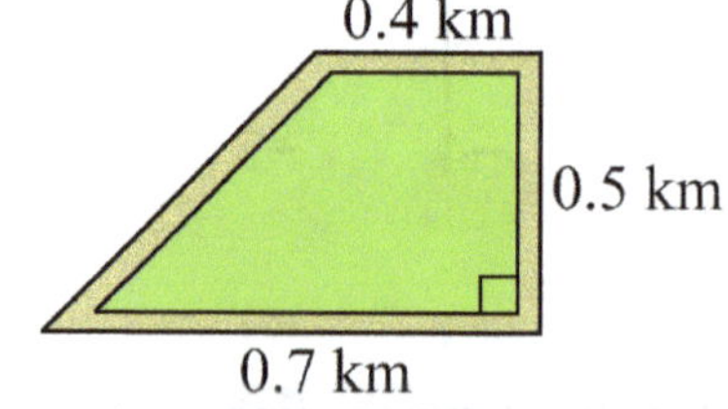

Mixed Review 9

1. Write an equivalent expression using the exponent laws, without negative exponents.

a. $3x^4y^5y^2 \cdot 6x^6 =$	**b.** $(3x)^{-3}$ =	**c.** $(3yz)^2$	**d.** $(b^{-2})^4$ =
e. $\dfrac{8x^5}{28x^8}$ =	**f.** $\dfrac{x^{-5}}{x^2}$ =	**g.** $\left(\dfrac{-2}{5y}\right)^2$ =	**h.** $\left(\dfrac{3s}{t^2}\right)^4$ =

2. Draw a dilation of triangle ABC...

a. from point A with scale factor 1/3

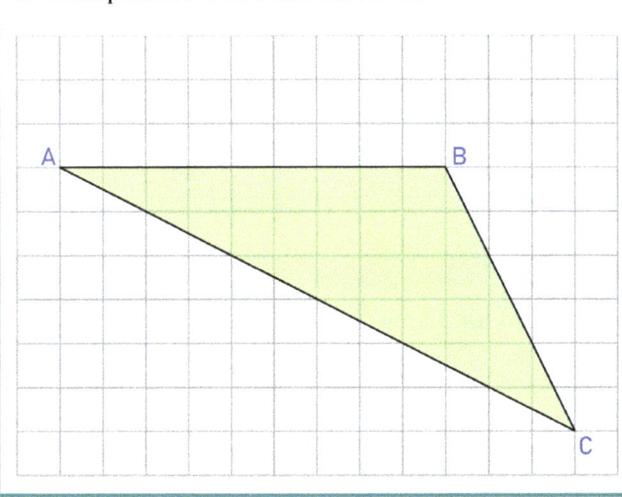

b. from point C with scale factor 1/2

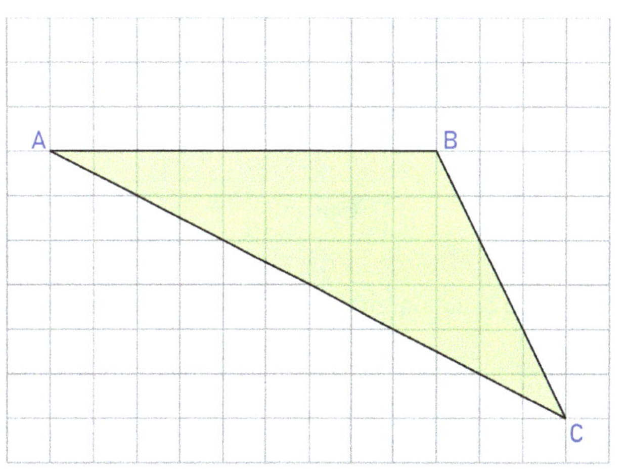

3. Eight tennis balls fit snugly in a cube-shaped container. Calculate what fraction of the total volume of the cube the tennis balls take up.
Hint: write this fraction using the formulas for the volumes, and simplify it.

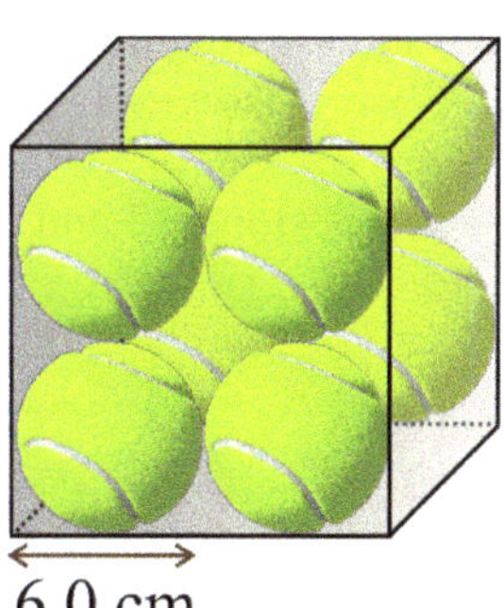

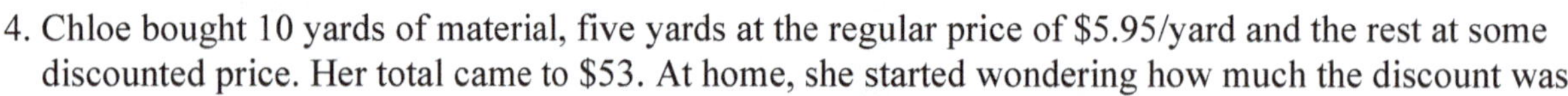

4. Chloe bought 10 yards of material, five yards at the regular price of \$5.95/yard and the rest at some discounted price. Her total came to \$53. At home, she started wondering how much the discount was. 

Write an equation to solve what the unknown discounted price was. Use p for the discounted price. Then solve your equation.

5. **a.** Make up two functions for the cost of renting a surfboard as a function of time. The first should be a proportional relationship, and the other nonlinear. Make your functions reasonable so that the cost of renting a surfboard for an entire day (8 hours) is \$50 at a maximum.

Give the linear function as an equation, and the nonlinear one as a table of values.

Function 1:

Function 2:

time (hours)	0	1	2	3	4	5	6	7	8
Cost (\$)									

b. Which function gives a better deal if you are renting a surfboard for 2 hours? For 6 hours?

6. Find the equation of each line, in slope-intercept form. Also graph the lines.

a. has slope −2 and passes through (−2, 6)

b. has slope 2/3 and passes through (4, −4)

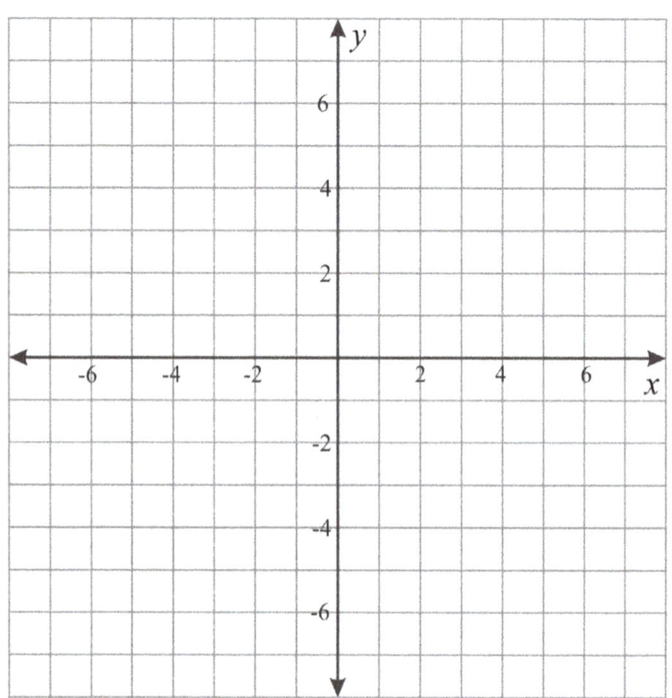

7. **a.** What is the equation of a horizontal line that passes through (−3, −5)?

b. What is the equation of a vertical line that passes through (9, 8)?

c. What is the equation of a line that is parallel to $y = 5x + 2$ and passes through (1, 2)?

8. Convert, rounding your answer to the same number of significant digits as the measurement. 

a. 71.0 cm = _________ in **b.** 2,400 kg = ____________ lb

c. 235 ft = __________ m **d.** 83.5 lb = __________ kg

e. 15.69 m = ___________ ft **f.** 4.5 in = __________ cm

1 inch = 2.54 cm

1 ft = 0.3048 m

1 kg = 2.2 lb

9. If one inch of rain falls over one square mile, how many raindrops fell? Give your answer in scientific notation, to four significant digits. 

Besides the well-known conversion factors, here are some facts you may need:

- One gallon = 231 cubic inches
- The size of raindrops varies but for this problem, use 10^5 raindrops per gallon.

Mixed Review 10

1. Solve.

a. $2 - \frac{x+1}{5} = x$	**b.** $4y + \frac{1-2y}{10} = 3y$
c. $\frac{x-10}{3} = \frac{3x+4}{5}$	**d.** $-x + \frac{7x-3}{4} = \frac{x}{2} - 2$

2. Rewrite the numbers in scientific notation correctly.

a. $72 \cdot 10^{-6}$

b. $233 \cdot 10^{6}$

c. $0.04 \cdot 10^{-2}$

d. $0.309 \cdot 10^{9}$

3. Compare the numbers, writing $<$, $>$, or $=$ in the box.

a. $1.9 \cdot 10^{8}$ ☐ $9.1 \cdot 10^{7}$

b. $9 \cdot 10^{-11}$ ☐ $9 \cdot 10^{-12}$

c. $5.2 \cdot 10^{-5}$ ☐ 0.000053

d. $9 \cdot 10^{-4}$ ☐ 0.0009

e. $7 \cdot 10^{4}$ ☐ 100,000

f. 0.00000053 ☐ $1 \cdot 10^{-6}$

4. Tell, without solving the equations, whether each equation has one unique solution, no solutions, or an infinite number of solutions.

a. $-3x + 6 = -3x - 9$

b. $-3x + 2 = -1$

c. $-3x + 4 = 4$

d. $-3x = 5 - 3x - 5$

5. Choose the relationships that are functions.

(1)

Input	**Height (cm)**	131	130	125	135	131
Output	**Name**	Henry	Harry	Helen	Hal	Holly

(2) Let S be a rule that takes any positive number x as input, and gives $1/x$ as output.

(3) Input is a person's social security number, output is their name.

(4) Input is a date (such as July 8), output is the person whose birthday it is.

6. Three functions are represented below.

a. Which one has the largest initial value?

b. Which one(s) are linear functions?

c. Which one has the smallest rate of change in the x-interval [6, 10]?

d. Describe each function in the x-interval [4, 6] as increasing, decreasing, or constant

Function 1:

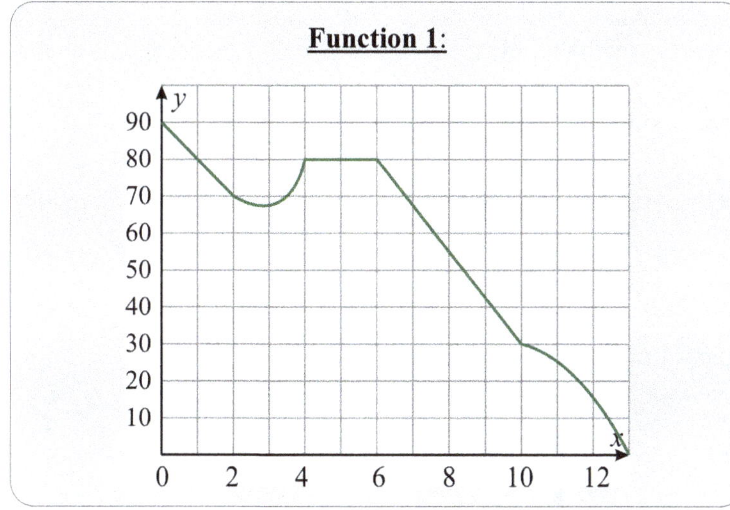

Function 2:

$y = 6.5x + 10$

Function 3:

x	y
0	80
2	65
4	50
6	35
8	20
10	5
12	−10

7. Grace deposits $5,000 in a savings account with a 6% interest. The interest will be added to the principal every year (compound interest). She is told that the final amount (A) on her account after t years is given by the formula $A = 5000(1.06)^t$. Is this function a linear function? Explain.

8. Construct two functions below by filling in the rest of the output values. In (a), the function should be a *linear* function. In (b), it should be *nonlinear*.

a.

Input (x)	−5	−4	−3	−2	−1	0	1	2	3	4	5
Output (y)					0	3	6				

b.

Input (x)	−5	−4	−3	−2	−1	0	1	2	3	4	5
Output (y)	63	56	49	42	35						

9. Find the volume of a square pyramid with a 30-cm by 30-cm bottom square, and a height of 24 cm.

10. Karen wants to do some remodeling in her kitchen. She is comparing the pricing from two companies. Company 1 charges $42 per work hour plus a fixed fee of $1,800 for the materials. Company 2 will charge $36 per work hour and $2,400 for the materials.

a. Write an equation for the cost (C) of using Company 1 as a function of work hours (h). Do the same for Company 2.

Company 1: ______________________________ Company 2: ______________________________

b. If it is estimated that the job will take 15 hours, which company is a better deal?

c. For what number of work hours does Company 1 cost the same as Company 2?

Systems of Linear Equations Review

1. Tell how many solutions each system of equations has by inspecting the equations. You do not have to find the solution(s).

a. $\begin{cases} y + 2x = -8 \\ 2y + 4x = -16 \end{cases}$

b. $\begin{cases} 7x - 2y = -1 \\ -2y + 7x = 3 \end{cases}$

c. $\begin{cases} x + y = -1/2 \\ 2y - 3x = 6 \end{cases}$

2. Solve the systems of equations depicted by the graphs.

a.

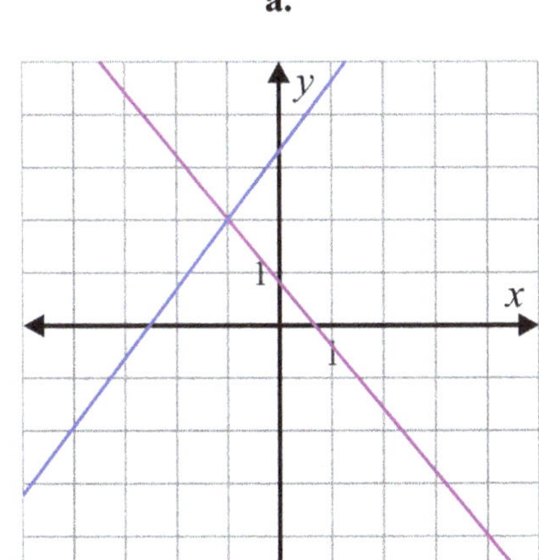

b.

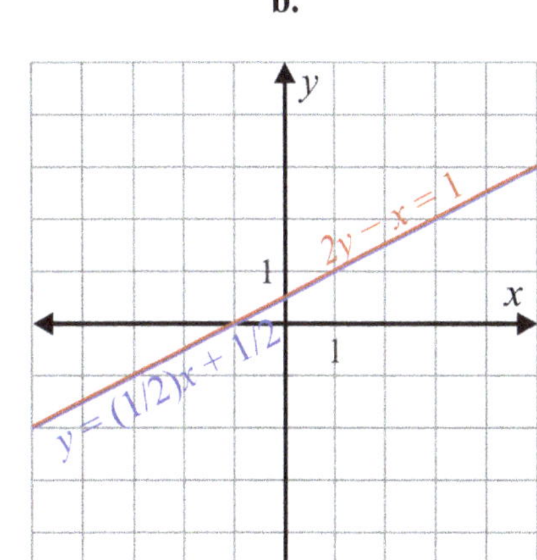

3. Find the system below that has a single solution, and solve it.

a. $\begin{cases} 4x - 2y = 1 \\ y = 2x + 6 \end{cases}$

b. $\begin{cases} y = (1/2)x + 6 \\ x + 6y = 0 \end{cases}$

c. $\begin{cases} 3(x + y) = 3 \\ y = -x + 1 \end{cases}$

4. Solve the system of equations by substitution. Then graph the lines. Verify that the intersection point of the lines is the solution you found algebraically.

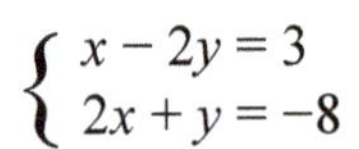

$$\begin{cases} x - 2y = 3 \\ 2x + y = -8 \end{cases}$$

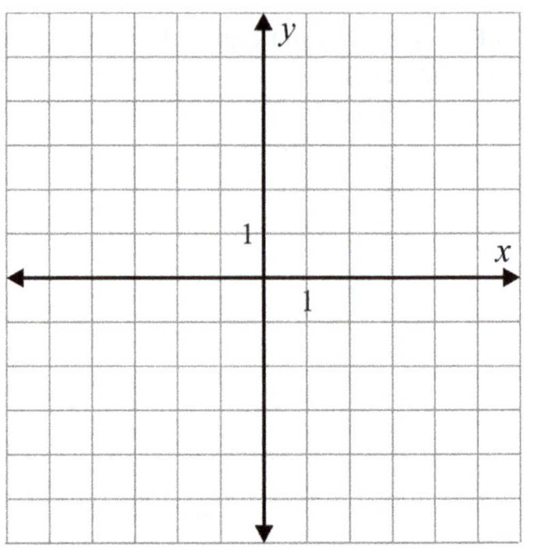

5. Solve each system of equations.

a. $\begin{cases} x = -(y - 5) \\ 2y = 12(1 - x) \end{cases}$

b. $\begin{cases} 4.5x - 3y = 0 \\ 7x + 2y = -2 \end{cases}$

c. $\begin{cases} 3x = -9(y + 1/3) \\ -2x - 6y = 2 \end{cases}$

d. $\begin{cases} 5x - 3(y + 2) = 0 \\ -5x + 6y = 7 \end{cases}$

6. Find the error in the solution of this system of equations. Then correct the error and solve the system.

$$\begin{array}{l} (1) \\ (2) \end{array} \left\{ \begin{array}{r} 6x + 3y = 3 \\ -2x + 9y = 11 \end{array} \right| \cdot 3$$

$\downarrow$

$$\left\{ \begin{array}{r} 6x + 3y = 3 \\ -6x + 27y = 33 \end{array} \right.$$

$$30y = 36$$

$$y = 5/6$$

$\downarrow$

$$\begin{aligned} (1)\quad 6x + 3(5/6) &= 3 \\ 6x + 5/2 &= 3 \\ 6x &= 3 - 5/2 = 1/2 \\ 6x &= 1/2 \\ x &= 1/12 \end{aligned}$$

However, the solution (1/12, 5/6) does *not* fulfill the 2nd equation:

$$\begin{aligned} -2(1/12) + 9(5/6) &\overset{?}{=} 11 \\ -1/6 + 45/6 &\overset{?}{=} 11 \\ 44/6 &\neq 11 \end{aligned}$$

7. Solve each system of equations and give the solutions as decimals, rounded to three decimal digits. Note: if you use a decimal in the intermediate steps, include at least five decimal digits. Round only the final answers to three decimals.

a. $\left\{ \begin{array}{l} 7x + 6y = -1 \\ 11x + 2y = 3 \end{array} \right.$

b. $\left\{ \begin{array}{l} 3.4x + 0.7y = 5 \\ 0.5x - 0.2y = -2 \end{array} \right.$

8. A bicycle shop sells both bikes and trikes. If there are 80 wheels and 31 vehicles, find the number of trikes the shop has.

9. Denny and Sammy were comparing their ages. Sammy said, "In four years, my age will be 2/3 of your age." Denny said, "And I am 14 years younger than double your age." Find their ages.

10. The digit sum of a two-digit number is 11. The number itself (its value) is two less than seven times its ones digit. Find the number.

11. A mixture of peanut butter and protein powder weighs 210 grams and contains 30% protein. The protein powder by itself contains 90% protein, and the peanut butter by itself contains 25% protein. Find how many grams of peanut butter and how many grams of protein powder are in the mixture.

12. Flying with the wind, a crow flies from his favorite tree to a nearby pond, a distance of 400 meters, in 48 seconds. Coming back, flying against the wind, it takes him 72 seconds. What is the crow's speed in still air? What is the wind speed?

Systems of Linear Equations Test

1. Tell how many solutions each system of equations has. (You do not have to find the solution(s).)

a. $\begin{cases} -y + 2x = 3 \\ 3y + 7x = -20 \end{cases}$ **b.** $\begin{cases} 3y - 4x = 6 \\ -6y + 8x = -12 \end{cases}$ **c.** $\begin{cases} 4x - 11y = -1/2 \\ -11y = 6 - 4x \end{cases}$

2. **a.** Give a value to s in such a manner that the system below has no solutions.

$$\begin{cases} y = 3x + 1 \\ y = sx - 2 \end{cases}$$

b. Graph the lines.

c. Explain how the graph shows that there are no solutions.

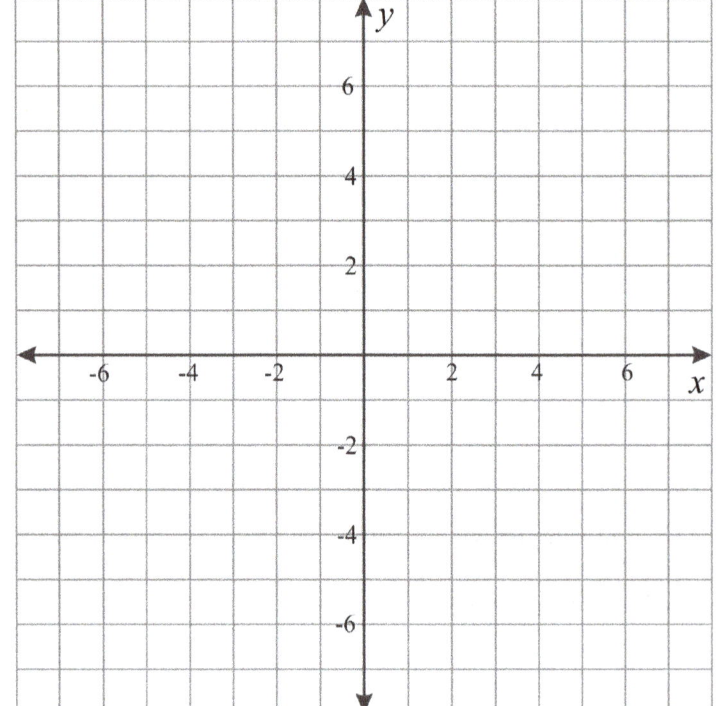

3. Solve each system of equations.

a. $\begin{cases} x = 10 - y \\ 5y = 7(x - 2) \end{cases}$

b. $\begin{cases} 3x - y = 14 \\ 6x + 3y = -12 \end{cases}$

4. Solve the system of equations using any method. Also, graph the lines.

$$\begin{cases} x + 3y = -2 \\ 2x - 5y = -15 \end{cases}$$

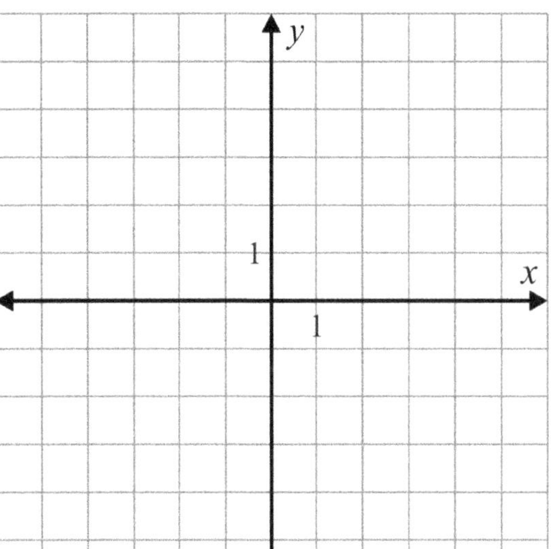

Solution: ________________

5. Solve the system of equations and give the solution as decimals, rounded to three decimal digits.

$$\begin{cases} x + 6y = -1 \\ y = 2.2x + 0.8 \end{cases}$$

6. Ann and Adam talked about their ages. Ann said, "In three years, I will be double your age." Adam said, "And seven years ago, you were 2.5 times my age." Find their ages now.

7. A restaurant has two kinds of tables: tables that seat 4 people, and tables that seat 6. If the restaurant has a total of 28 tables, and those can seat 132 people, how many tables that seat 6 do they have?

8. Train 1 leaves Station A, heading for Station B, a distance of 70 km, at 3 PM and travels with a constant speed of 90 km/h. Train 2 leaves Station B, heading for Station A and travels at the constant speed of 105 km/h, at the same time.

At what time do they meet?

What distance has the second train traveled by that time?

	distance	***velocity***	***time***
Train 1			
Train 2			

Mixed Review 11

1. Show that the two trapezoids are similar by describing a sequence of transformations that could map trapezoid ABCD to the smaller trapezoid.

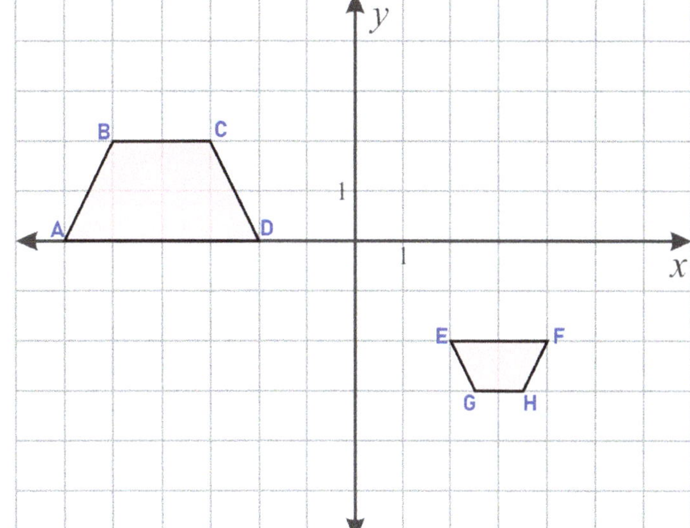

2. Solve the equations.

a. $s + 2 - 5s = 3 - 8(s - 1) - 6s$	**b.** $6 + \dfrac{2x - 5}{3} = x - 2$
c. $0 = \dfrac{x - 5}{3} + \dfrac{x + 5}{4}$	**d.** $10 = y + \dfrac{2 - 3y}{5}$

3. Give an example of each type of equation that has $2x - 5$ on the left side of the equation.

a. No solutions:

$2x - 5 =$

b. One solution:

$2x - 5 =$

c. An infinite number of solutions:

$2x - 5 =$

4. Find the equation of each line.

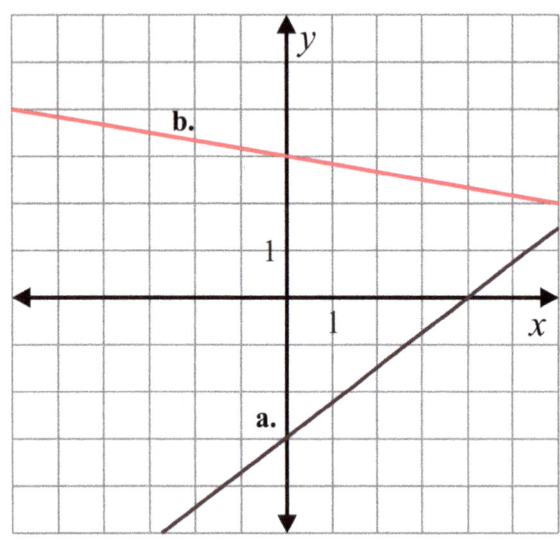

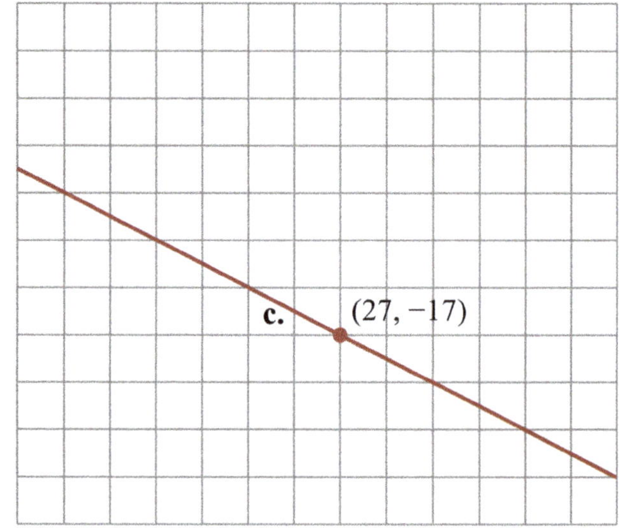

a.

c.

b.

5. Solve each system of equations.

a. $\begin{cases} y = -2(x + 10) \\ 2(y - x) = 8 \end{cases}$	**b.** $\begin{cases} 2x - 4y = -3 \\ 5x + 3y = 1 \end{cases}$

6. **a.** Two dogs, Rocky and Charlie, start racing at the same time. Rocky runs at a steady speed of 10 m/s. After 20 seconds, they are 8 m apart from each other, Rocky leading. How fast is Charlie running?

	distance	*velocity*	*time*
Rocky			
Charlie			

b. What would Charlie's speed need to be, so that after 20 seconds, he would only be 2 m behind Rocky?

7. Brass is an alloy of copper and zinc. There exist many different kinds of brasses where copper and zinc are in different proportions. If you have the two alloys listed below on hand, how much of each would you need to melt together in order to produce 5,000 kg of brass that contains 75% copper and 25% zinc?

	copper	*zinc*
Alloy 1	90%	10%
Alloy 2	65%	35%

8. An elephant runs (or “fast-walks”) at a constant speed of 6 m/s.

a. Convert this speed to meters per minute.

b. Write an equation for the distance (d) the elephant covers as a function of time in *minutes* (t).

c. Plot your equation. Design the scaling of the vertical axis so that the point for three minutes fits on the graph.

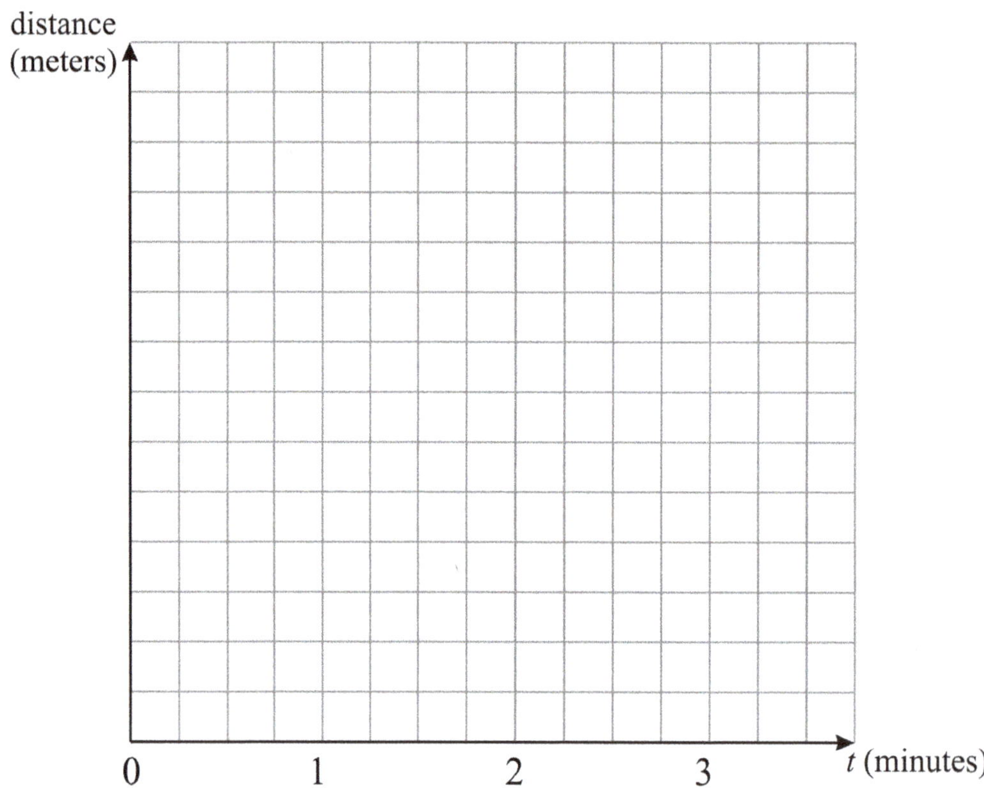

d. What is the slope of the line in your graph?

e. The table below shows the distance and time when Elizabeth bicycles at a constant speed.

Time (min)	2	4	5	10	15
Distance (m)	800	1600	2000	4000	6000

Who goes faster, the elephant or Elizabeth?

f. How much larger distance does the faster one cover in 3 minutes?

Mixed Review 12

1. Write in order from the smallest to greatest value: $2 \cdot 10^{-5}$ $\quad -2 \cdot 10^{5}$ $\quad 0.0005$ $\quad -2 \cdot 10^{-5}$ $\quad 2 \cdot 10^{5}$

2. Are the triangles similar? Explain how you know.

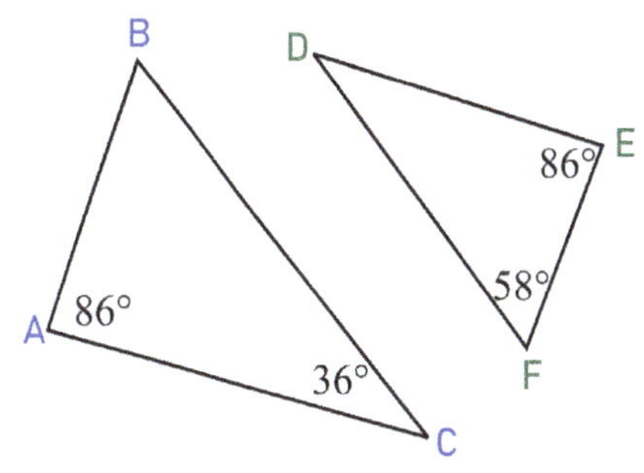

3. The table shows the cost of renting a snowboard as a function of time.

Days	Cost ($)
0	10
1	65
2	120
3	175
4	230
5	285

a. Write an equation to model this (linear) relationship.

b. What is the rate of change of this function?

What does that signify in this context?

c. What is the initial value of this function?

What does that signify in this context?

4. Sketch a graph to depict the distance of the boat from the shore as a function of time.

The boat starts out far away from the shore. It approaches the shore at a steady speed for two days. Then it advances towards the shore at a greater speed for one day, and reaches it. It stays docked for two days. Then it goes out to the ocean again, its distance from the shore growing in a linear manner, for three days. After that, its distance from the shore continues to increase, but gradually slows down. For day 10, it stays in its place.

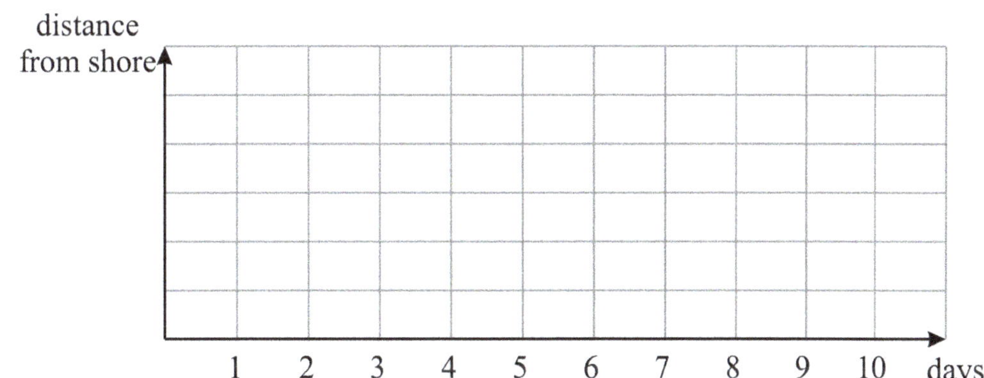

5. The perimeter of an isosceles triangle is 65 units. If its base side was 7 units longer, the triangle would be equilateral. Find the sides of the isosceles triangle.

6. ABCD is a parallelogram. Find the value of x.

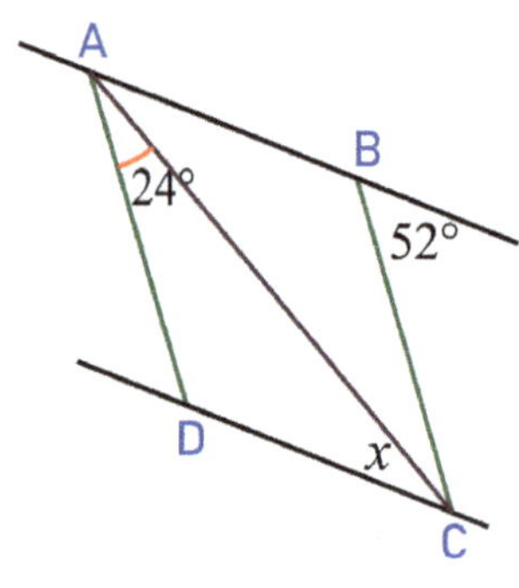

7. **a.** Draw a line with a slope of 3/2 that goes through the point (0, −2). Then write its equation.

b. Draw a line with a slope of −3 that goes through the point (−4, 5). Then write its equation.

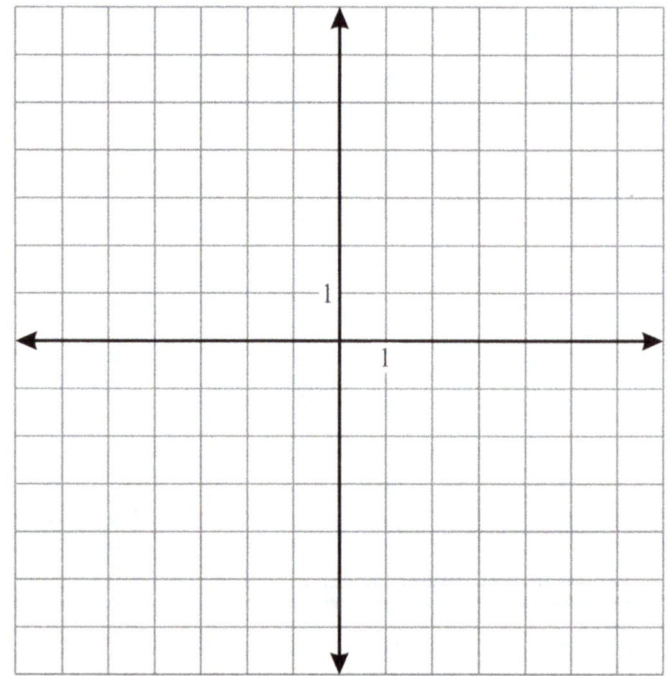

8. Match the descriptions and the equations.

a. Passes through (4, 1) and has slope 1/4	**(i)** $y = (1/4)x - 2$
b. Passes through (−1/4, 1/4) and has slope 0	**(ii)** $x = 1/4$
c. Passes through (1/4, 1/4) and has no slope	**(iii)** $y = (1/4)x$
d. Passes through (4, −1) and (−8, −4)	**(iv)** $y = 1/4$

9. For each line given in standard form, find its x and y-intercepts. Then graph the line.

a. $4x - 2y = 8$

b. $9x + 3y = -18$

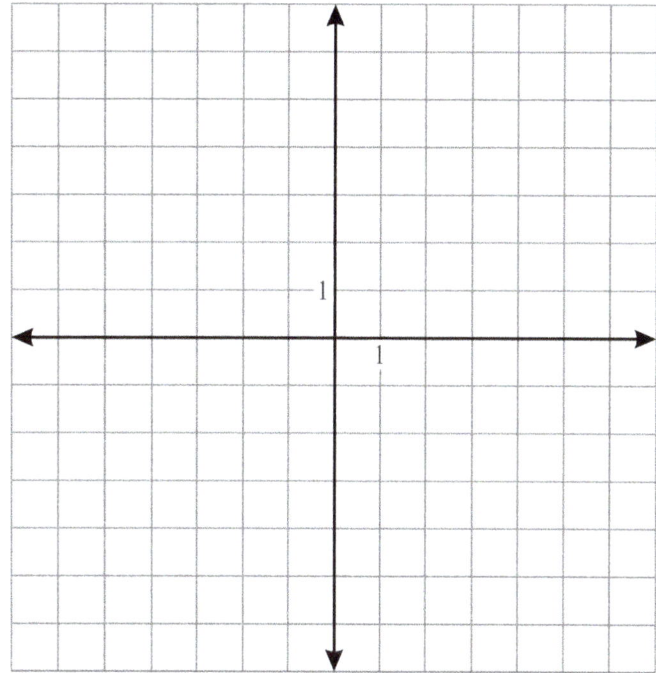

10. An item had three price increases: for 3%, for 4%, and for 9%. Now it costs \$268.55. What was its price before these increases?

11. Often, drinking glasses are in the form of a cut cone. Calculate the volume of the glass on the right, using this technique:

(1) Calculate the volume of the entire cone shape that is 25 cm tall.
(2) Calculate the volume of the cut part, that is also a cone (at the bottom).
(3) Subtract the above two results.

(The image is not to scale.)

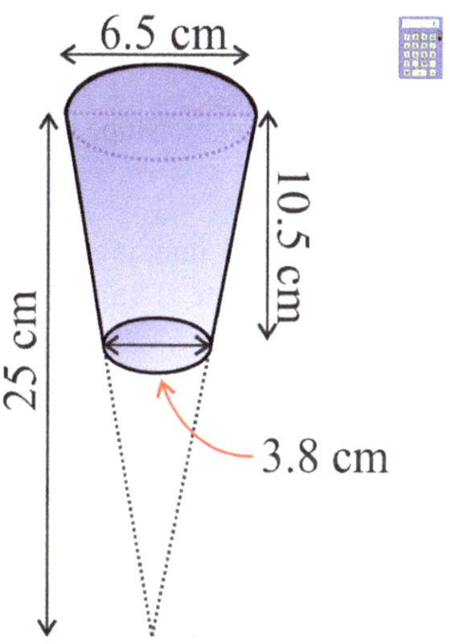

Bivariate Data Review

1. The scatter plot below shows the life expectancy at birth and the birth rate per 1000 population for almost 200 countries in the world.

   a. Describe the pattern of association you see in the scatter plot.

   b. Describe any special features visible in the scatter plot.

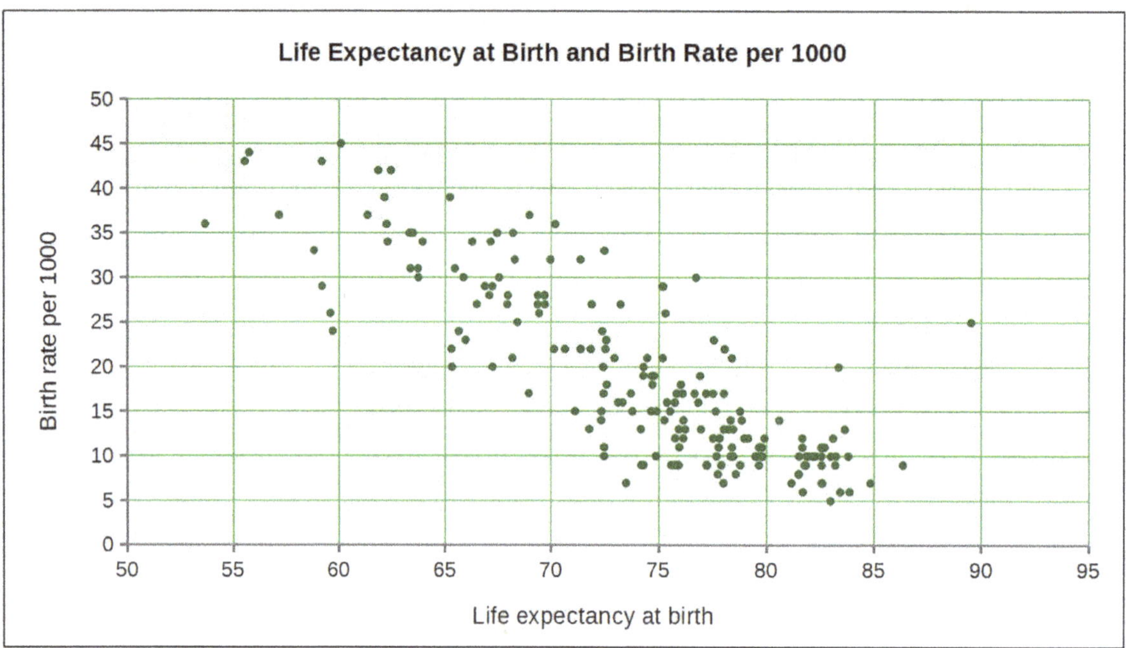

   c. Draw a line to represent the trend in the data.

   d. Find the (approximate) equation of your line.

   e. What does the slope of your equation signify in this context?

   f. Using your line, predict the life expectancy at birth in a country where the birth rate is 25 per 1000 population.

   g. Using your line, predict the birth rate per 1000 population in a country where the life expectancy at birth is 55 years.

   h. The dot at (89.5, 25) is Monaco. What is the difference between the predicted life expectancy at birth for a country with a birth rate of 25 per 1000, and the actual life expectancy of Monaco?

2. The scatter plot below has to do with the cost of airplane tickets and the distance covered by the flight. A spreadsheet program calculates the equation of the trendline to be $C = 0.111d + 76.9$ where C is the cost (in dollars) and d is the distance (in miles).

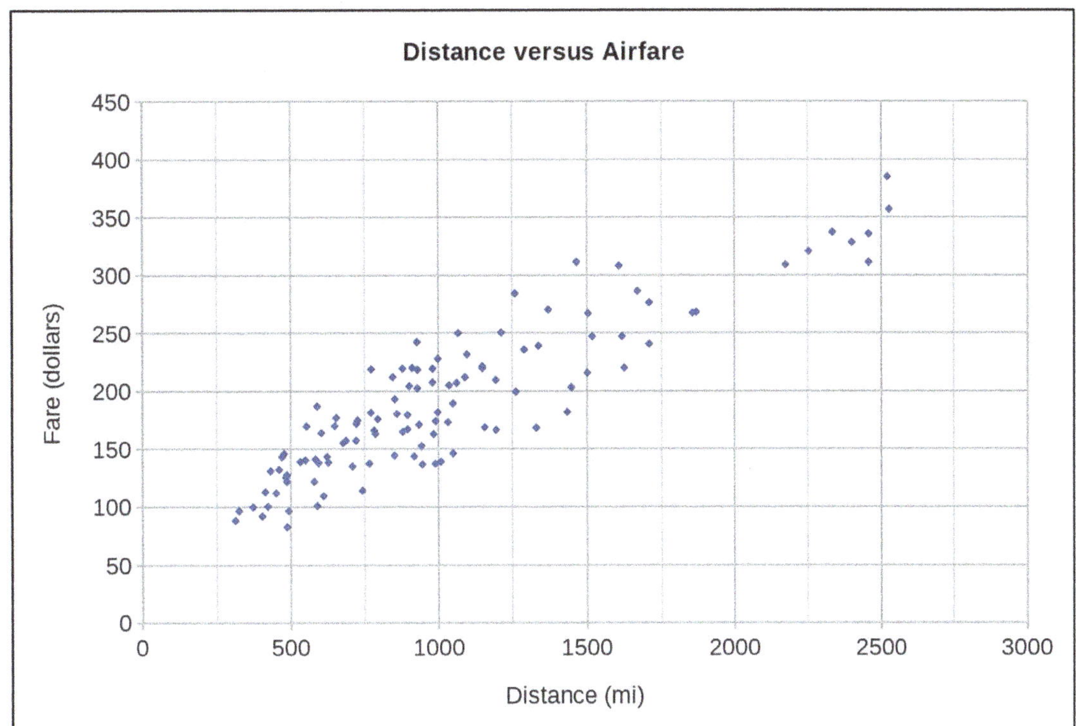

a. Sketch this line in the image.

b. Find four different points in the image for tickets costing slightly over 300 dollars. What are the distances covered in those flights?

c. What is the distance predicted by the trendline equation, if a ticket costs 310 dollars?

d. What is the ticket cost predicted by the trendline equation, if the flight distance is 800 miles?

e. For each 100-mile increase in distance, how much would you expect the cost of the ticket to increase?

f. Based on the trend line, if you look at the airfares of \$300 and \$250, what is the associated difference in distance, to the nearest mile?

3. A multinational company asked its employees what was their favorite sport (to watch or to play). The results are categorized based on the location of the employees.

a. Calculate the totals. Then calculate the relative frequencies, to the nearest percent, based on the row totals, and write them in the table on the right.

	soccer	baseball	basketball	Total
Asia	17	1	5	
Europe	85	7	60	
Middle East	40	3	10	
North America	11	45	26	
Total				

	soccer	baseball	basketball	Total
Asia				100%
Europe				
Middle East				
North America				

b. Make at least three observations about the data. For one example, we can see that people in Middle East prefer soccer far above the other two sports.

c. Is there an association between a person's location and their preferred sport? Explain your reasoning.

d. If you randomly select a person who prefers soccer, how likely is it that the person is from Europe?

e. If you randomly select a person who prefers basketball, how likely is it that the person is from the Middle East?

f. If you randomly select a person from Asia, how likely is it that the person prefers basketball?

4. Which of the following numbers of 2, 9, or 24, would you put in the empty box so that...

a. there is no association between the variables?

b. girls are far more likely to play an instrument than boys?

c. boys are far more likely to play an instrument than girls?

	plays an instrument?	
	yes	no
Boys	12	18
Girls	6	

Bivariate Data Test

1. James started a training program to improve his 5-kilometer running time. The plot below shows the number of days he has been doing the program versus his running time in minutes.

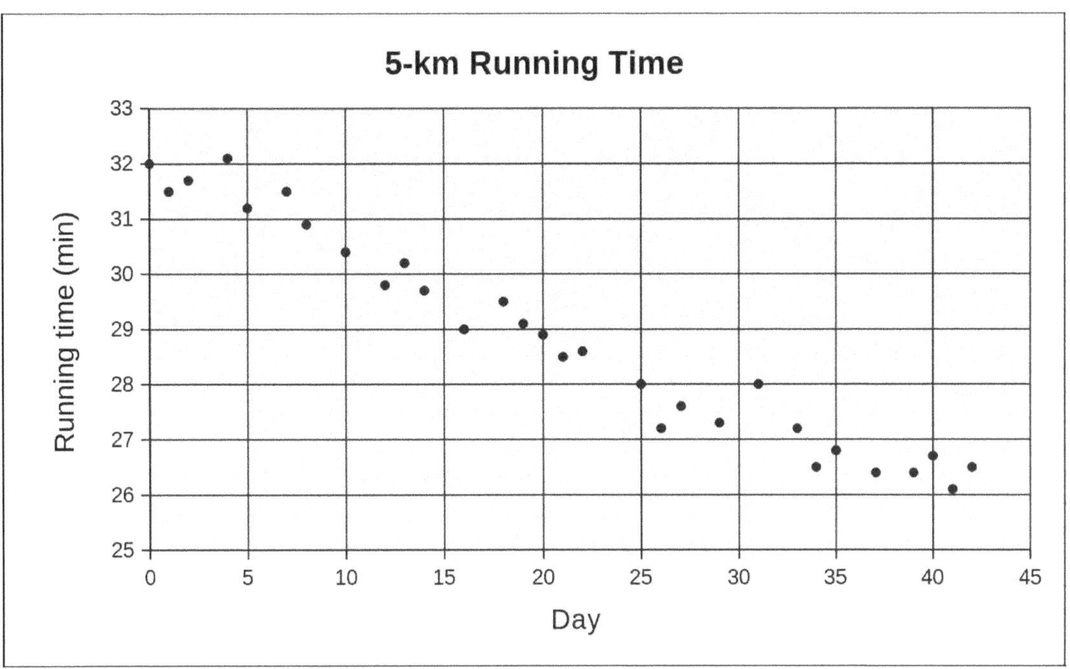

a. Is there a relationship or an association between the two variables? If so, what kind?

b. Which of the following points would be considered an outlier, if it was added to the plot?

(24, 28.3), (24, 29), (24, 31.3)

c. Draw a trend line to fit the data.

d. Find the (approximate) equation of your line.

e. What running time does your equation predict for day 24?

f. What does the slope of your equation signify in this context?

g. What does the y-intercept of your equation signify in this context?

2. The scatter plot below shows the diameter of pizzas from various pizzerias versus their price.

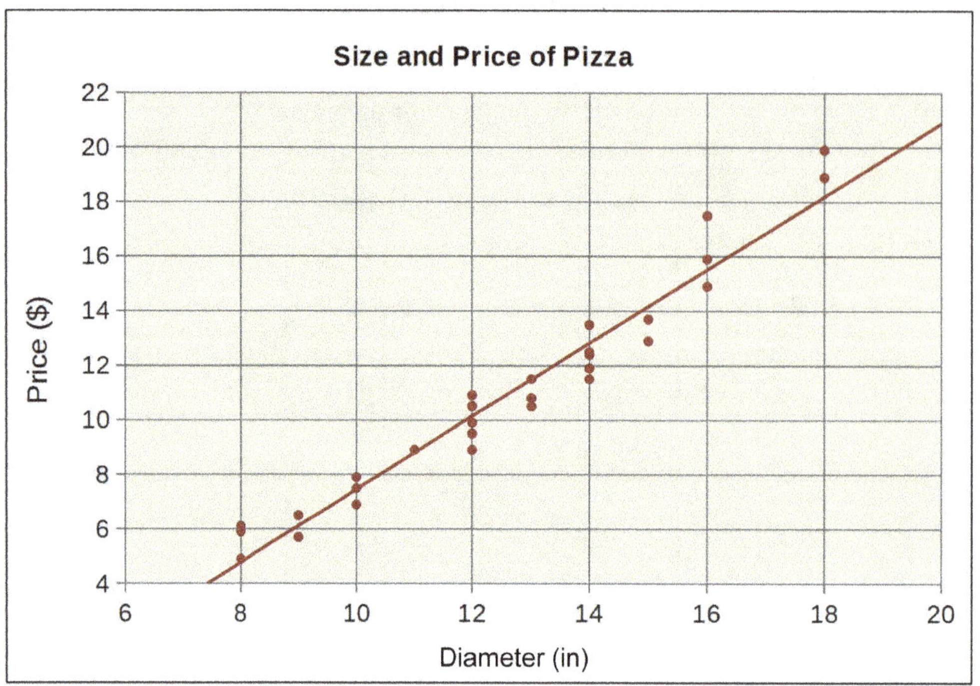

a. The equation of the trend line is $y = 1.34x - 5.95$.
What does the slope of this equation signify in this context?

b. Can we extrapolate and use the equation to predict the cost of, say, a pizza with 1-inch diameter? Explain why or why not.

c. Using the equation, predict what the price would be for a pizza with a 15-inch diameter.

d. Using the equation, predict the diameter of a pizza that costs $15.

3. The two-way table shows how many students in a certain high school didn't take foreign language, or took Spanish, French, German, or some other foreign language.

	None	Spanish	French	German	Other	**Totals**
Male	117	20	4	3	1	**145**
Female	97	40	9	3	2	**151**
Totals	**214**	**60**	**13**	**6**	**3**	**296**

a. Is there an association between the variables? Justify your answer.

b. If you choose randomly a student who took French, what is the chance they're female? Give your answer to the nearest percent.

c. If you choose randomly a male student, what is the chance they didn't take any foreign language? Give your answer to the nearest percent.

4. **a.** Find the two-way table with **no association** between the variables.

b. Explain how you found it.

	Exercise frequency per week				
Daily hours on social media	**Never**	**Once**	**Twice**	**Thrice**	**4x or more**
1	4	5	10	13	11
2	9	10	7	7	5
3-4	19	10	9	3	2
5-6	24	13	5	2	0

	Ate fish in the previous week?	
	Yes	**No**
Australia	23	20
Japan	87	2
Portugal	73	14
United States	56	47
Brazil	16	68

	Has two or more siblings	
	Yes	**No**
14-15 years	13	28
16-17 years	6	12
18-19 years	14	27

	Dropped out of school	
	Yes	**No**
high school	98	673
vocational school	93	367

Mixed Review 13

1. Lines l and m intersect at point A. Find the measure of angle x.

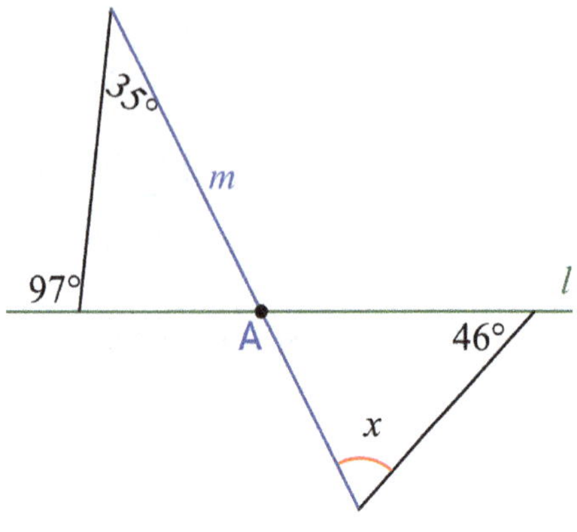

2. Find the (principal) square roots.

a. $\sqrt{10^2 - 6^2}$ b. $\sqrt{29 + 7}$ c. $\sqrt{\frac{81}{16}}$

d. $\sqrt{0.16}$ e. $\sqrt{0.01}$ f. $\sqrt{1.21}$

3. Solve the equations. Give each solution rounded to three decimals.

a. $18 + t^3 = 134$	b. $5x^2 = 220$	c. $109 - v^3 = 11$

4. Find the statements that are in error, and correct them.

a. $\frac{79}{123}$ is rational because it is a fraction (a whole number divided by a whole number).

b. $\sqrt{121}$ is irrational because it has a square root.

c. 3π is irrational because π is irrational, and an irrational number multiplied by a rational number (3) is irrational.

d. 0.19191919... is irrational because it is an unending decimal.

5. The diagonal of a rectangle measures 5.5 m and its one side is 2.80 m. How long is the other side?

6. This is a triangular prism (a tent-shaped structure).
Find the length marked with x.

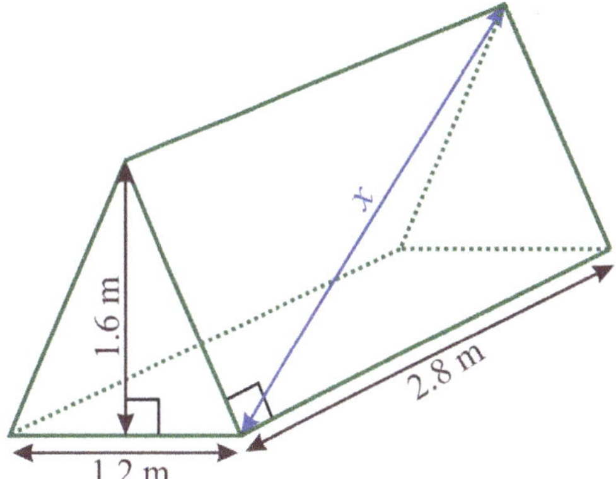

7. Graph the volume of a water reservoir as a function of time to match the description below.

Your graph will be a rough sketch only, without a specific scale on the vertical axis.

From June onward, the level of water was decreasing in a linear manner, over the summer months. During September, the level of water increased continually, quickly at first, and then at a slower and slower pace, because of rains. For the month of October, there was no change in the water level. The level then increased continually in a linear manner through November and December. In January, there again was no change.

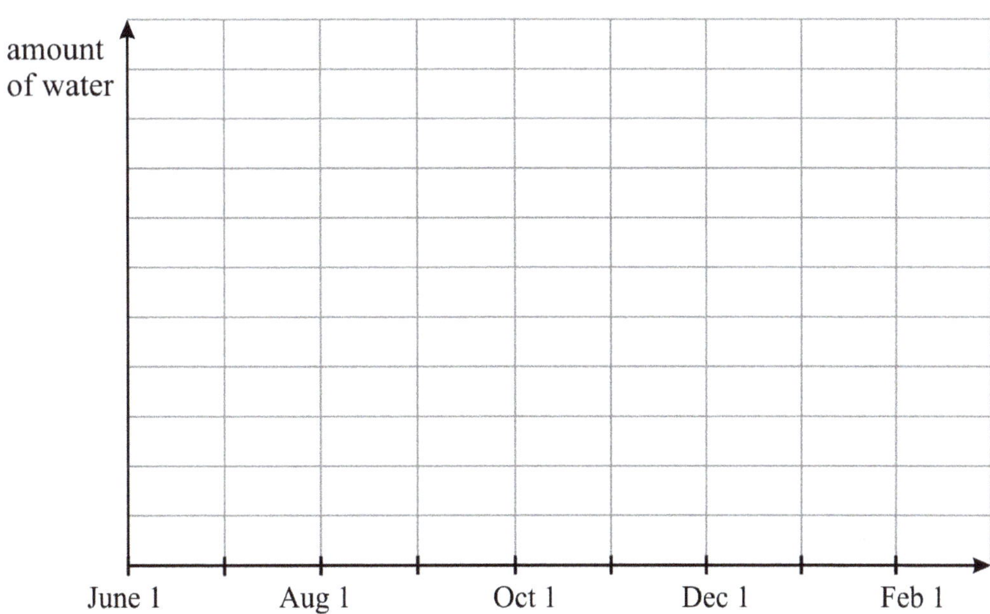

8. **a.** Line L is plotted on the right. Line M is parallel to line L, and passes through the point (3, 0).

Find the equation of Line M, in slope-intercept form, and plot it.

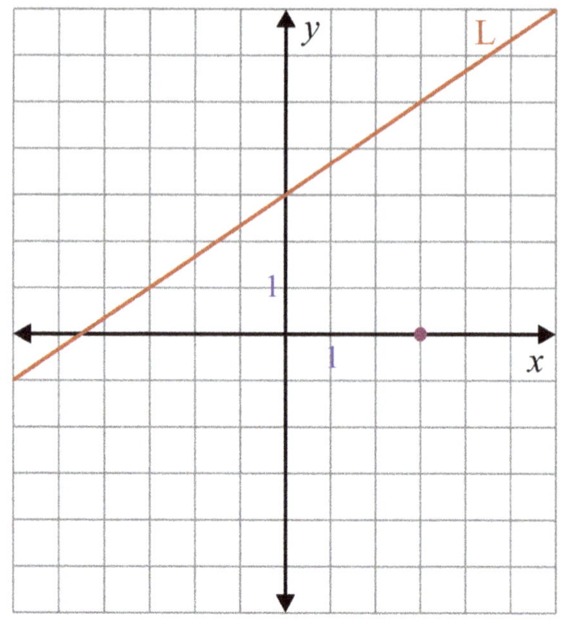

b. Line N is perpendicular to line L, and passes through the point (3, 0).

Find the equation of Line N, in slope-intercept form, and plot it.

9. Solve each system of equations.

a. $\begin{cases} 5x - 7y = 3 \\ -6x + 8y = 1 \end{cases}$	**b.** $\begin{cases} 3x = -2(y + 1) \\ -x - 3y = 0 \end{cases}$

10. A group of chickens and cows has 42 heads and 100 legs. How many chickens and how many cows are there?

11. The scatter plot shows the age and the resting heart rate of a group of males who use a fitness wearable.

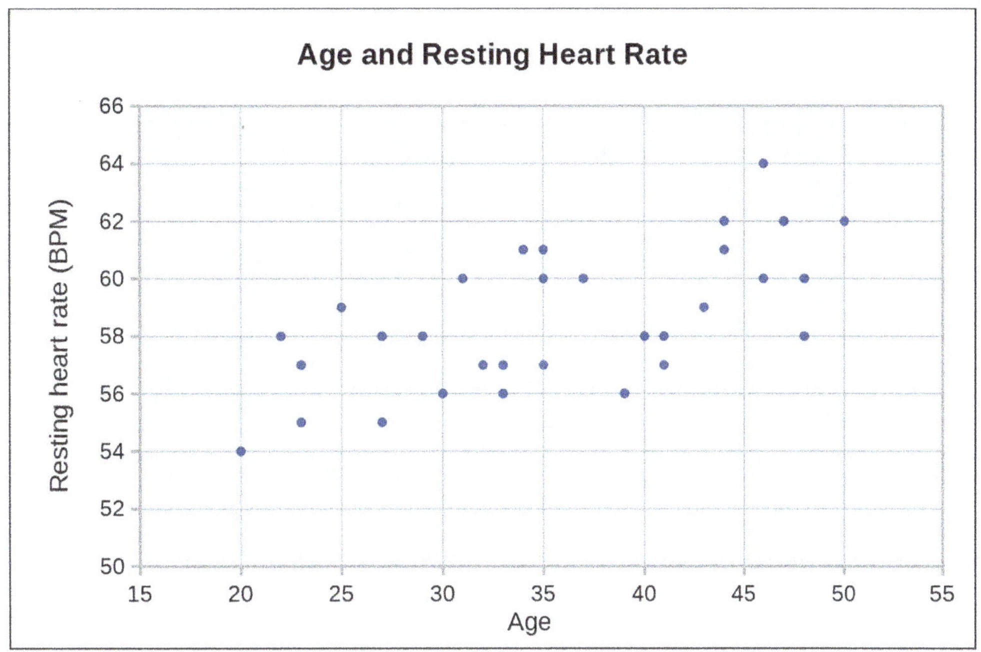

a. Describe the general trend and any special features in the scatter plot.

b. Draw a trend line on the graph.

c. Find the equation of your line.

d. According to your equation, when a person's age increases by one year, what change in the resting heart rate is associated with that?

e. What heart rate does your equation predict for a 30-year-old male who wears a fitness wearable?

f. What heart rate does your equation predict for a newborn?
(A normal heart rate for newborns is from 120 to 160 beats per minute.)

What does that tell us about extrapolating with this data?

12. Claire owns a salon. She conducted a survey among her clientele concerning how often they had a haircut, on average. The results are shown in the table below.

	Male	Female	Total
every month	56	15	71
every two months	31	37	68
every three months	17	46	63
every four months	8	25	33
TOTALS	112	123	235

Relative frequencies:

	Male	Female
every month		
every two months		
every three months		
every four months		
TOTALS	100%	100%

a. Which statement(s) are correct?

(i) Among those who get a haircut every three months, most of them are female.

(ii) If you randomly choose a male client from Claire's clientele, most likely they get a haircut every two months.

(iii) Most of the clients who get a haircut every month are males.

b. Calculate the relative frequencies based on the *column* totals, to the nearest percent.

c. Is there an association between the variables? Explain.

Mixed Review 14

A basic calculator (not a graphing calculator) is allowed for all the exercises.

1. The points A(5, −1), B(3, −4), and C(1, −3) are the vertices of a triangle. It is first reflected in the x-axis. Then it is dilated from the origin with a scale factor of 2. Lastly, it is rotated 90° clockwise around the origin. What are the coordinates of its vertices now?

2. **a.** How many solutions does this equation have?

   b. Modify only one term in the equation so that the resulting equation will have no solutions.

$$10s - 6 = 2s$$

3. Solve the systems of equations.

a. $\begin{cases} 3x - y = 14 \\ 6x + 3y = -12 \end{cases}$	**b.** $\begin{cases} 3x - 8(y - 2) = 0 \\ 2x = 8y + 1 \end{cases}$

4. Some of the statements below are incorrect. Correct them.

a. $\sqrt{121}$ is irrational because it is a square root.

b. 0.831831831 is rational because its decimal expansion repeats.

c. $\sqrt{13}$ is irrational because it is a square root of a non-perfect square.

d. $\frac{\pi}{6}$ is irrational because π is irrational, and an irrational number divided by a rational one is irrational.

e. $\frac{\sqrt{63}}{4}$ is rational because it is a fraction.

5. How much milk that is 3% butterfat and cream that is 15% butterfat should you mix in order to get one liter of "milky cream" that is 12% butterfat?

Filling the chart can help.

	volume	***butterfat percentage***	***butterfat amount***
3% milk			
15% cream			
Mixture			

6. Find the error in the solution of this system of equations. Then correct the error and solve the system.

$$\begin{array}{l} (1) \\ (2) \end{array} \begin{cases} 3x - 7y = 4 \\ -5x + y = -4 \end{cases} \Bigg| \cdot 7$$

$$\downarrow$$

$$\begin{cases} 3x - 7y = 4 \\ -35x + 7y = -28 \end{cases}$$

$$-32y = -24$$

$$y = 3/4$$

$$\downarrow$$

$$\begin{aligned} (1) \quad 3x - 7(\mathbf{3/4}) &= 4 \\ 3x - 21/4 &= 4 \\ 3x &= 4 + 21/4 = 37/4 \\ 3x &= 37/4 \\ x &= 37/12 \end{aligned}$$

However, (37/12, 3/4) does *not* fulfill the 2nd equation.

7. Solve for the unknown side. In (a), give your answer to one decimal. In (b), give the exact answer. 

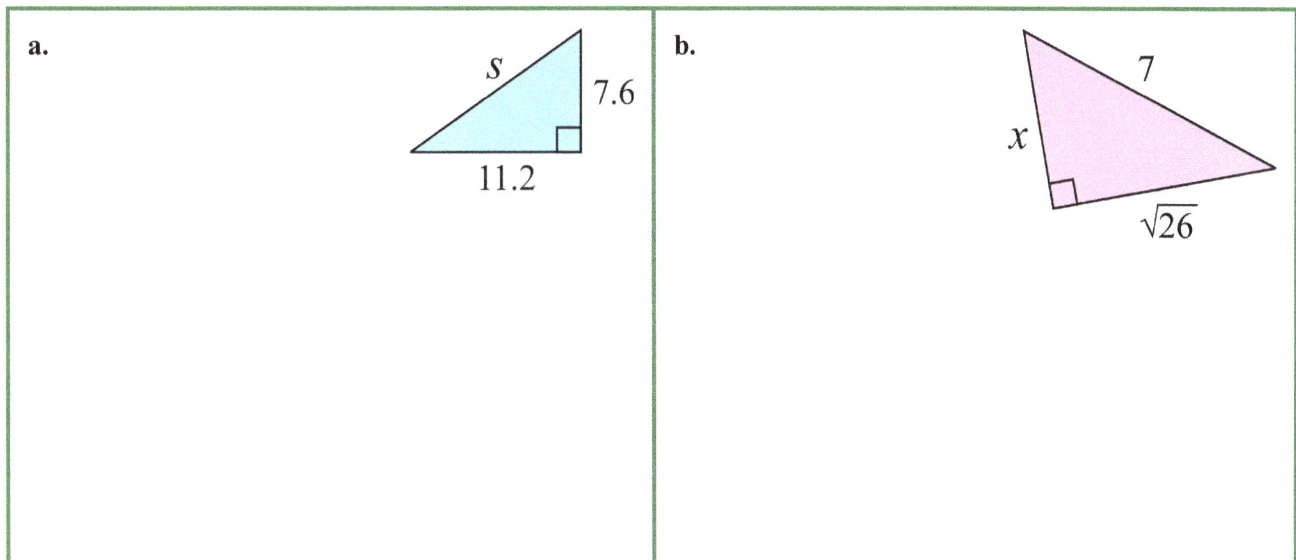

8. Calculate the area of an equilateral triangle with 16.5-inch sides.
Don't forget to draw a sketch. 

9. **a.** Fill in the missing numbers in this two-way table.

b. Is there an association between the two variables?

Favorite hobbies of 2nd graders

	Boys	**Girls**	**Total**
Sports	26	21	47
Music	12	13	
Reading		20	33
Arts & Crafts	5	13	18
Video games	16		23
Cooking	2	5	7
Photography	6		
Total	80	84	164

10. The volume of a circular prism is $V = \pi r^2 h$, where r is the radius of the circle and h is its height. Solve this for r.

11. **a.** Draw a line to fit the trend in the scatter plot below.

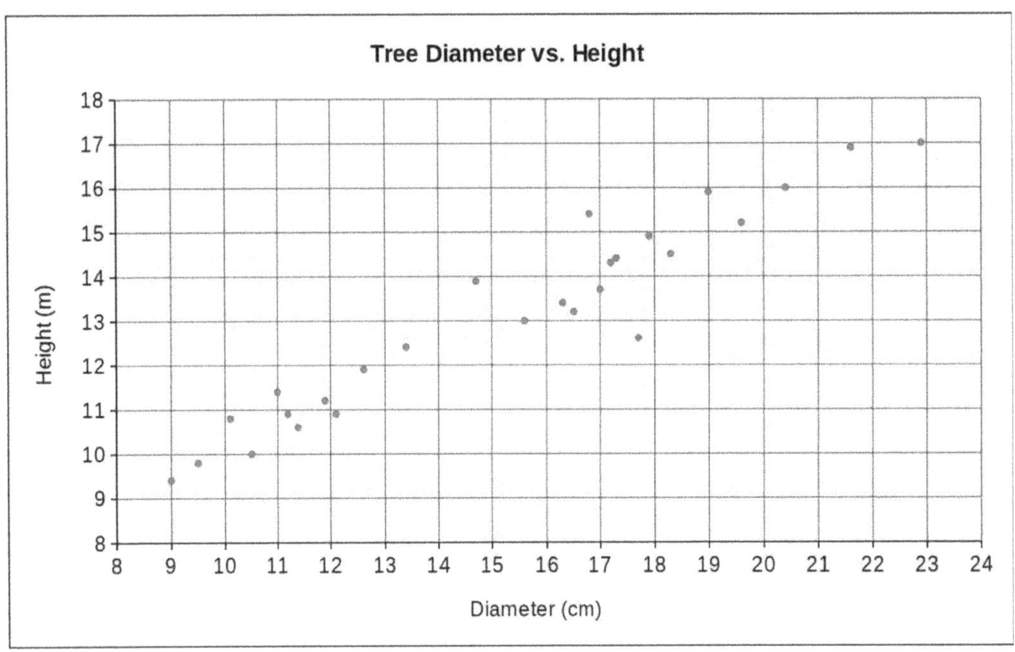

b. Find the (approximate) equation of your line.

c. What does the slope of your equation signify in this context?

d. What does the y-intercept of your equation signify in this context?

e. Using your equation, predict the diameter of a tree that is 16.5 m tall.

End-of-Year Test — Grade 8

This test is quite long, because it contains questions on all of the major topics covered in *Math Mammoth Grade 8*. Its main purpose is to be a diagnostic test: to find out what the student knows and does not know of the topics covered in the curriculum. Since the curriculum follows the Common Core Standards for 8th grade, the test questions cover most of those standards.

Since the test is fairly long, I don't recommend that you have the student do it in one sitting. Feel free to break it into 2-5 parts and administer them on consecutive days, or perhaps in a morning/evening/morning/evening. Use your judgment.

Important: **A calculator is *not* allowed in the first two sections: Exponents and Scientific Notation, and Irrational Numbers. A basic calculator (not a graphing calculator) *is* allowed for the rest of the test.** The questions where calculator usage is most appropriate show a little calculator picture on the right.

The test is evaluating the student's ability in the following content areas:

- exponent laws
- scientific notation, including in calculations
- irrational numbers
- simple equations involving square or cube root
- geometric transformations, including dilations
- basic angle relationships
- volume of spheres and cylinders
- solving linear equations
- determining the number of solutions to a linear equation
- word problems that involve a linear equation
- concept of a function
- concept of a linear function
- rate of change and initial value of a function
- describing functions
- slope and graphing linear equations
- using the Pythagorean Theorem in mathematical and real-world problems
- solving systems of two linear equations
- solving word problems that lead to a system of two linear equations
- associations in scatter plots
- fitting a trend line to a scatter plot; interpreting a trend line for a scatter plot
- associations in two-way tables

Use your judgment in grading. You can give points or partial points for partial answers.

Question #	Max. points	Student score
Exponents and Scientific Notation		
1	8 points	
2	9 points	
3	4 points	
4	2 points	
5	2 points	
	subtotal	/ 25
Irrational Numbers		
6	5 points	
7	5 points	
8	3 points	
9	2 points	
	subtotal	/ 15
Geometry		
10	3 points	
11	2 points	
12	3 points	
13	2 points	
14a	3 points	
14b	3 points	
15	3 points	
16	3 points	
	subtotal	/ 22
Linear Equations		
17	4 points	
18	4 points	
19	6 points	
20	2 points	
21	2 points	
22	3 points	
	subtotal	/21
Functions		
23	2 points	
24a	1 point	
24b	2 points	
24c	2 points	

Question #	Max. points	Student score
Functions		
25a	1 point	
25b	1 point	
25c	1 point	
25d	1 point	
25e	1 point	
26a	2 points	
26b	1 point	
26c	1 point	
26d	1 point	
	subtotal	/17
Graphing Linear Equations		
27a	1 point	
27b	1 point	
27c	2 points	
28	3 points	
29	3 points	
30	3 points	
	subtotal	/13
The Pythagorean Theorem		
31	4 points	
32	3 points	
33	3 points	
	subtotal	/10
Systems of Linear Equations		
34	6 points	
35	3 points	
36	3 points	
37	3 points	
	subtotal	/15
Bivariate Data		
38	3 points	
39	3 points	
40	3 points	
41	5 points	
	subtotal	/14
	TOTAL	/152

Grade 8 End-of-Year Test

Instructions to the student:
Answer each question in the space provided. When applicable, round your answers to a reasonable accuracy according to the context of the problem. A calculator is *not* allowed in the first two sections of the test. A basic calculator (not a graphing calculator) *is* allowed for the rest of the test.

Exponents and Scientific Notation (no calculator allowed)

1. Find the value of the expressions.

a. $-2^4 =$	**b.** $(-2)^4 =$	**c.** $7^{-2} =$	**d.** $6^3 \cdot 6^8 \cdot 6^{-9} =$
e. $31 \cdot 10^{-3} =$	**f.** $10^5 + 10^4 =$	**g.** $\left(\frac{-2}{3}\right)^3 =$	**h.** $\frac{4^{10}}{4^7} =$

2. Write an equivalent expression using the exponent rules, and without negative exponents.

a. $(-2s)^3 =$	**b.** $(12x)^2 =$	**c.** $(y^3)^5 =$
d. $2x^6 \cdot (-3x^2) =$	**e.** $(y^{-3})^2 =$	**f.** $(4v)^{-2} =$
g. $\left(\frac{7x}{3y}\right)^2 =$	**h.** $\left(\frac{-x^2}{5x}\right)^3 =$	**i.** $\left(\frac{3b}{c^5}\right)^4 =$

3. Write the numbers in scientific notation.

a. 193,000,000

b. 3,080,500,000,000

c. 0.00046

d. 0.0000009

4. The earth's mass is $6.0 \cdot 10^{24}$ kg. Neptune's mass is $1.0 \cdot 10^{26}$ kg.
What fraction is the earth's mass of Neptune's mass?

5. One gold atom weighs about $3.3 \cdot 10^{-22}$ grams.
How many gold atoms are in 99 g of gold?

Irrational Numbers (no calculator allowed)

6. Plot the following numbers *approximately* on the number line. Do not use a calculator, but think about between which two whole numbers the root lies, and whether it is close to one of those whole numbers.

a. $-2\sqrt{2}$ **b.** $\sqrt{82}/3$ **c.** $-\sqrt{10}$ **d.** $\sqrt{35} - 1$ **e.** $\sqrt{8} + 3$

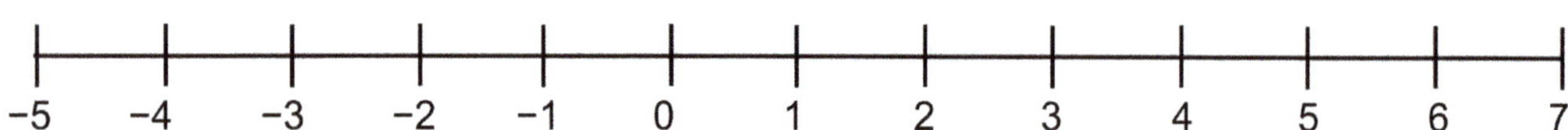

7. Place the numbers in the correct places in the diagram of *real numbers* = the set of both rational and irrational numbers. Note: the set of whole numbers is {0, 1, 2, 3, 4, 5, ...}.

$9, -109, \frac{13}{8}, \sqrt{3}, \sqrt{49}, 2\pi, 7.89, 0.4\overline{1}, \frac{\sqrt{400}}{8}, 0, -\frac{4}{9}, \sqrt{2} + 1, -\frac{35}{7}, \frac{5}{\sqrt{11}}, \sqrt{900}$

Real Numbers

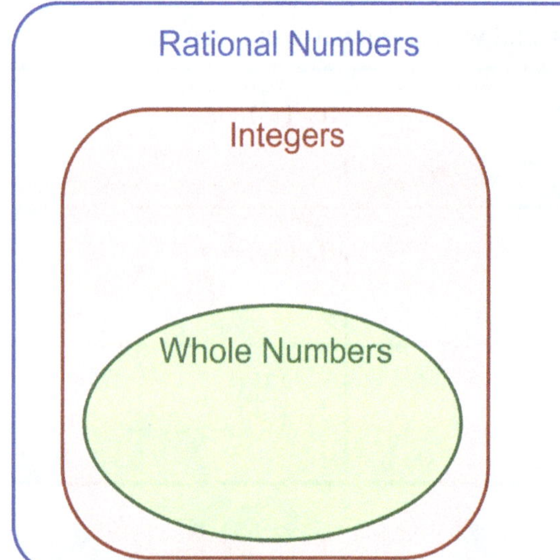

Irrational Numbers

8. Solve. If the answer(s) are not rational, give them in root form.

a. $x^2 = 54$	**b.** $3n^2 = 147$	**c.** $z^3 = 64$

9. Write the repeating decimal $0.\overline{71}$ as a fraction.

A basic calculator (NOT a graphing calculator) is allowed from this point onward.

Geometry

10. Show that the two triangles are similar by describing a sequence of transformations that maps triangle ABC to the smaller triangle.

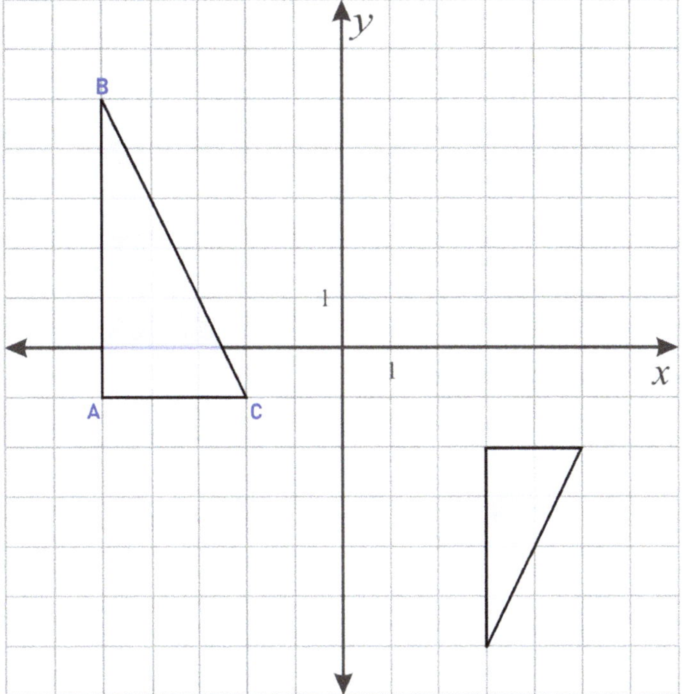

11. Explain a sequence of transformations that can map trapezoid ABCD to the smaller trapezoid.

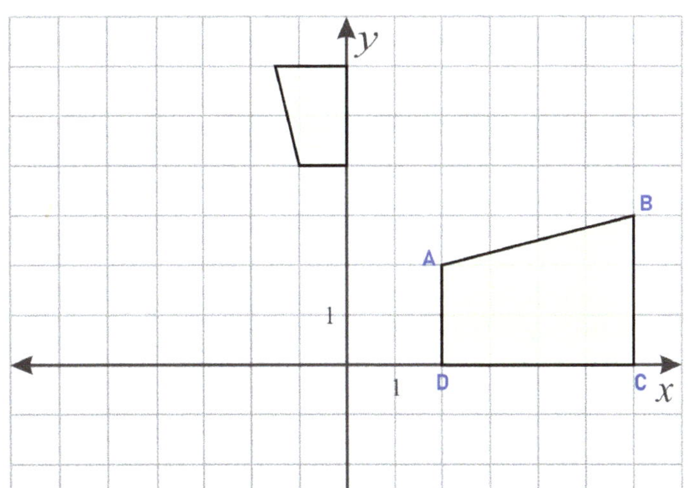

12. Triangle ABC has the vertices A(−2, 5), B(−5, 4) and C(−3, 0). It is transformed in the following ways: First, it is reflected in the y-axis. Then it is translated two units down and one to the left. Lastly, it is rotated 90° clockwise around the origin.

What are its coordinates after the transformations?

13. Find the value of x, and show your work.
(You can add additional labels to the image, to be able to reference different parts.)

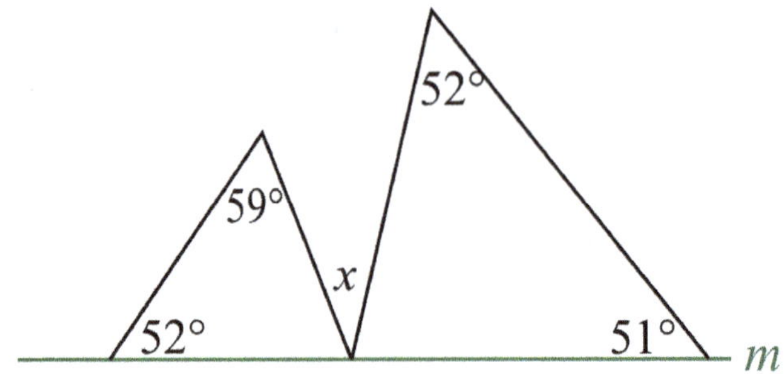

14. **a.** Find the value of x.

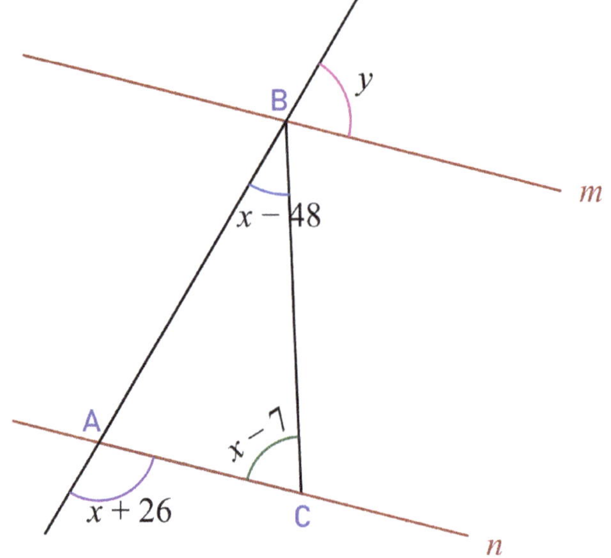

b. Lines m and n are parallel. Find the value of y.

15. Margaret has a spherical glass vase where she keeps flowers. Its diameter is 6.0 inches. How much water does she need to fill it 2/3 full? Give your answer both in cubic inches and in cups. *Note:* 1 cup = 14.4375 cubic inches. 

16. Lucas is designing a coffee cup in the shape of a circular cylinder. If its interior diameter is 6.2 cm, what should its (inner) height be, so that its volume would be 340 ml? *Note:* 1 ml = 1 cubic centimeter. 

Linear Equations

17. Solve.

a. $10s + 8 = 7s - 2(s - 5)$	b. $20 - 3(x + 4) = 14 - 5x$

18. Solve the equations.

a. $\frac{2x - 3}{5} - x = 2$	b. $\frac{y - 3}{4} = \frac{1 - y}{5}$

19. Solve the equations. Indicate whether each equation has one, none, or an infinite number of solutions.

a. $6x - 1 = 6(x - 1)$	b. $-5x + 1 = 6(x - 1) - 5$	c. $6x - 12 = 6(x - 2)$

20. Landon bought 4,500 concrete blocks at $1.35 apiece, but he got a discount on a third of them. If his total came to $5,775, find how much the discount was.

21. The sum of four consecutive whole numbers is 2,342. What are the numbers?

22. The price of an item is reduced by 27%, and then a 6% sales tax is added. You pay $34.82. What was the original price of the item?

Functions

23. **a.** Explain why the graph on the right does *not* depict a function.

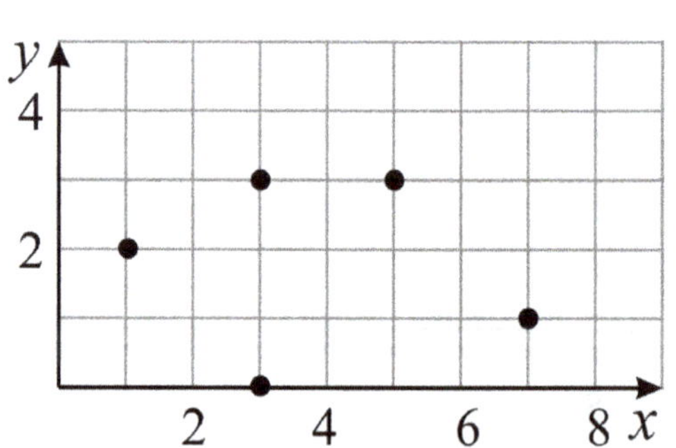

b. Place one of the numbers of 3, 6, or 9 to the empty space in the table so that it is a function.

Input	Output
2	3
5	9
7	5
3	1
	3
9	6

24. Farms A and B both grow strawberries and allow customers to get them for a lower price if they pick them themselves. Farm A charges the customer using the graph below. Farm B uses the table.

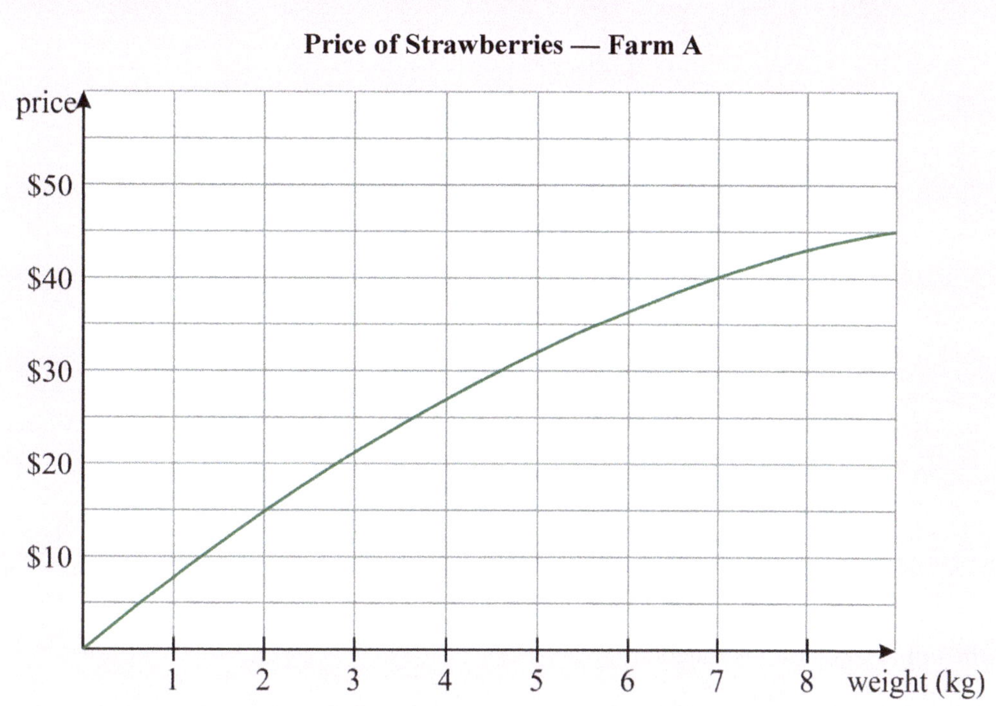

Price of Strawberries — Farm B

Weight (w)	Cost (C)
1 kg	$6.25
2 kg	$12.50
3 kg	$18.75
4 kg	$25.00
5 kg	$31.25
6 kg	$37.50
7 kg	$43.75

a. Which of the two functions is linear? Write an equation for it, using C for cost and w for weight.

b. Find the rate of change for each function from $w = 2$ to $w = 3$.

c. Which farm provides the better deal if you pick 4 kg of strawberries? If you pick 7 kg?

25. The graph below shows the cost for horse riding on a farm as a function of time.

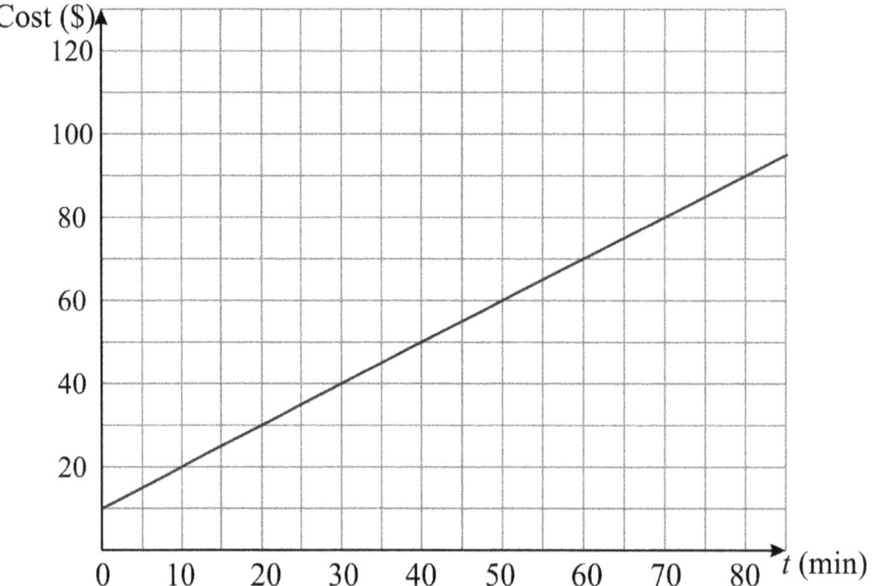

a. What is the initial value of this function?

b. What does it signify in this context?

c. What is its rate of change?

d. What does it signify in this context?

e. Write an equation for the graph.

26. **a.** Describe the function depicted in the graph below by intervals of x-values as increasing, decreasing, or constant, and also as linear or nonlinear.

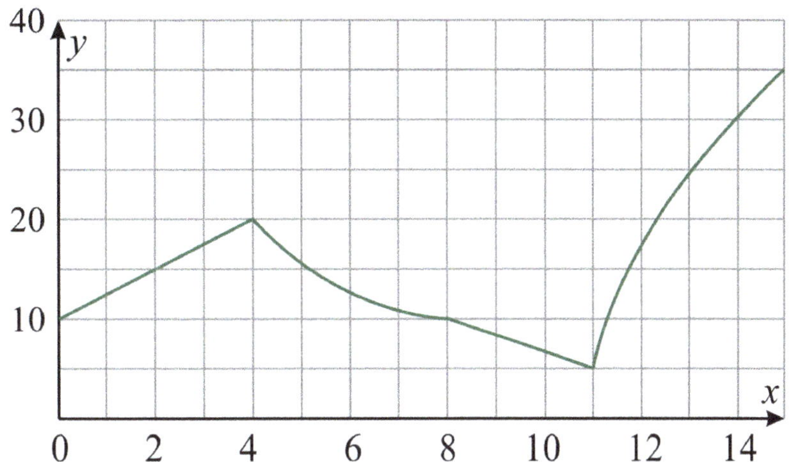

From $x =$ 0 to $x =$ 4 : ____________________

From $x =$ ____ to $x =$ _____: ____________________

From $x =$ ____ to $x =$ _____: ____________________

From $x =$ ____ to $x =$ _____: ____________________

b. Graph in the same grid a linear function that passes through the points (4, 25) and (8, 20).

c. Find its equation.

d. List the rates of change for both functions from $x = 8$ to $x = 11$.

Graphing Linear Equations

27. Find the equation of each line, in slope-intercept form:

a. has slope −2/3 and passes through (0, 4)

b. is horizontal and passes through (2, −3)

c. has slope 5 and goes through the point (6, 5).

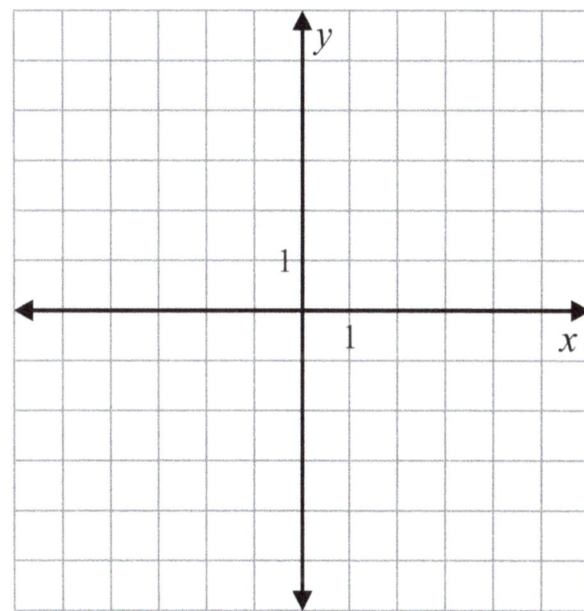

28. Refrigerator companies make estimates of how much energy their fridges consume in typical usage. The table shows how many kilowatt-hours (kWh) of energy Fridge 1 consumed over time, and the graph shows the same for Fridge 2.

Fridge 1

time (mo)	energy (kWh)
3	120
6	240
9	360
12	480

Fridge 2

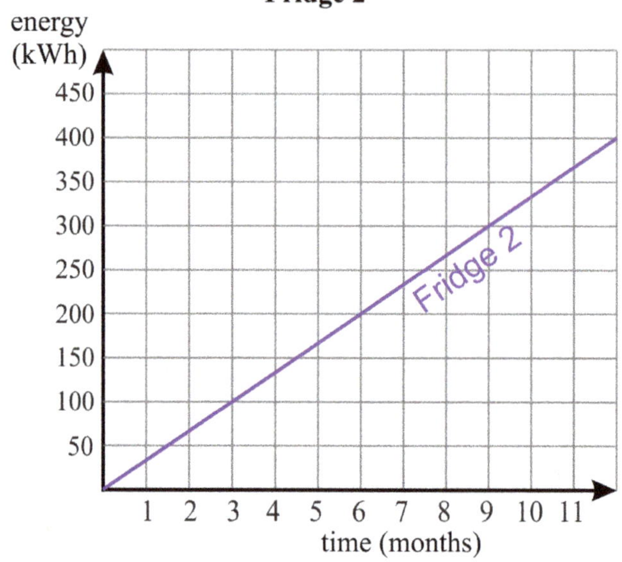

a. Which fridge consumes more electricity in *three* months?

How much more?

b. Write an equation for each fridge's energy consumption, relating the energy (E, in kWh) and the time (t, in months).

c. Plot the equation for Fridge 1 in the grid.

29. Graph the lines.

a. $y = (-1/2)x - 3$

b. $2x - 3y = 6$

c. $x = -4$

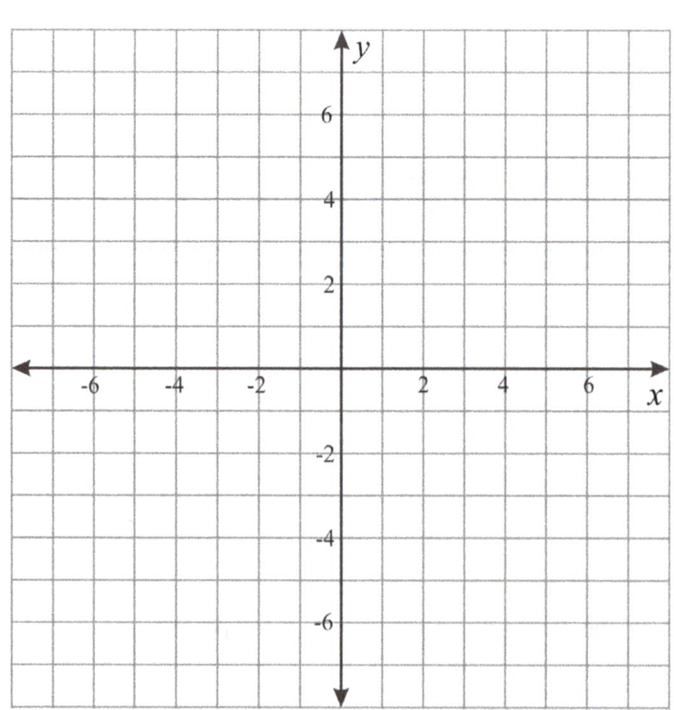

30. Find a so that the point $(a, 2)$ will fall on the same line as the points $(3, 14)$ and $(-7, -6)$.

The Pythagorean Theorem

31. Solve for the unknown side.

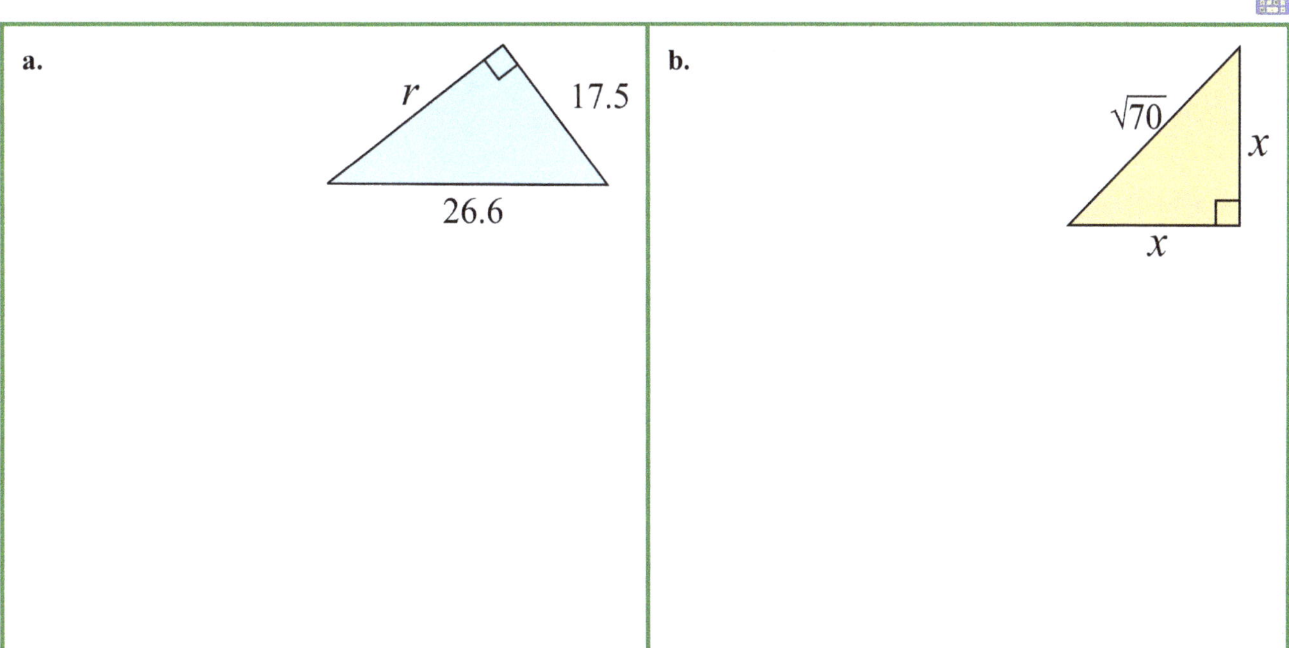

32. How long is the rafter (the roof piece) in this chicken coop design? Give your answer in feet and inches.

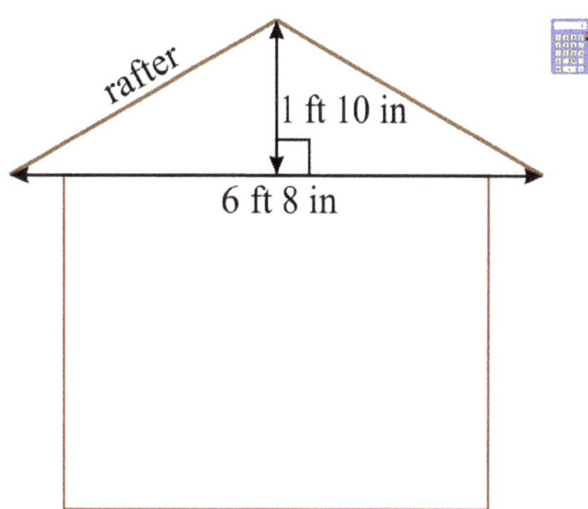

33. **a.** Find the height of this square pyramid.

b. Find its volume.

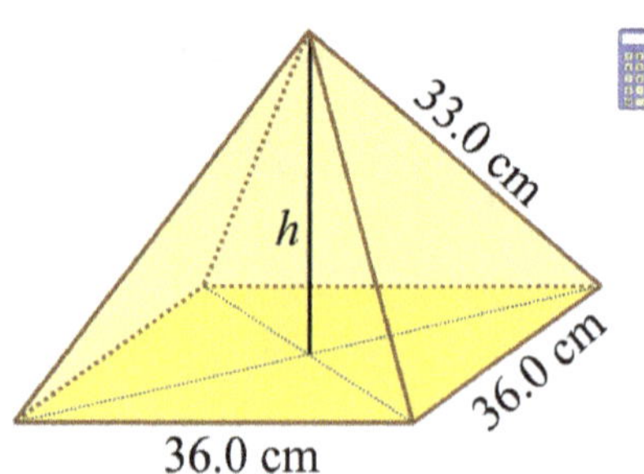

Systems of Linear Equations

34. Solve each system of equations. Give the solutions as fractions.

a. $\begin{cases} 2x - 3y = 8 \\ 3x + 4y = -5 \end{cases}$	**b.** $\begin{cases} -x = 4(y + 5) \\ 2x = -12y - 10 \end{cases}$

35. Tell how many solutions each system of equations has by inspecting the equations. You do not have to find the solution(s).

a. $\begin{cases} 4x - 2y = -8 \\ 2y - 4x = 1 \end{cases}$

b. $\begin{cases} 4x - 2y = -1 \\ -4y + 2x = 2 \end{cases}$

c. $\begin{cases} 4x - 2y = -1 \\ 2y - 4x = 1 \end{cases}$

36. A restaurant has two kinds of tables: ones that seat 4, and ones that seat 6. They have a total of 106 tables (some are in storage), and can seat 500 people.

Write a system of two equations to model this situation, and solve it.

37. Greta said to Susan, “In ten years, my age will be 3/4 of your age.” If the sum of their ages is 127, find their ages now.

Bivariate Data

38. Explain whether each scatterplot shows an association between the variables. If yes, classify the association as linear or nonlinear, increasing or decreasing.

a.

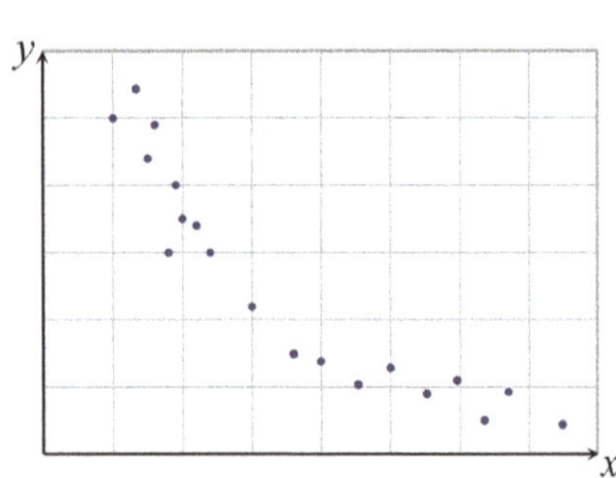

b.

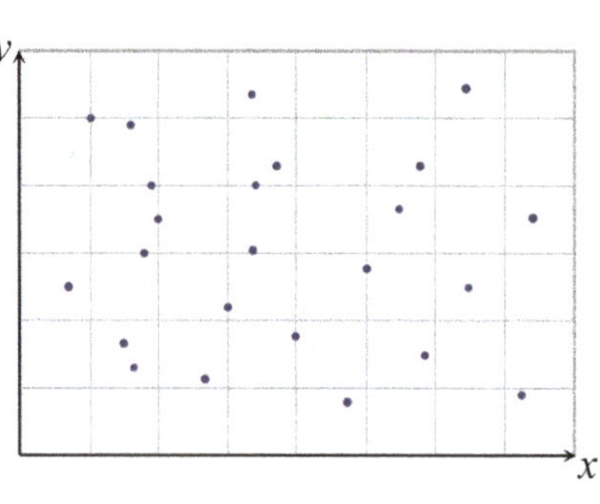

c.

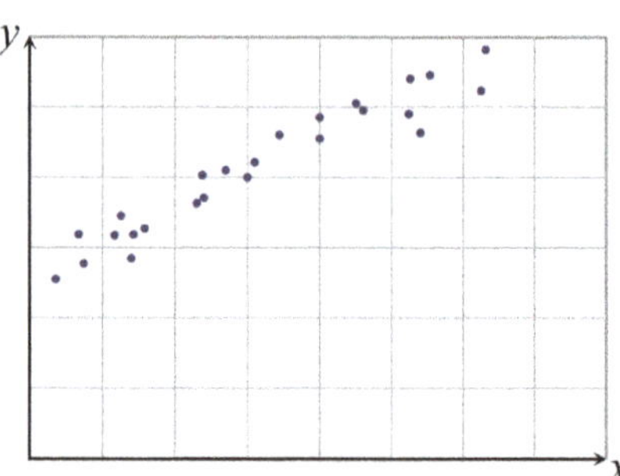

39. Is there an association between the two variables? Also, explain how you know that.

Age	Exercises	Does not exercise	Total
15-24	51	47	98
25-34	52	52	104
35-44	52	47	99
45-54	28	24	52
55-64	22	27	49
TOTAL	205	197	402

40. Which of the following numbers of 3, 6, or 24, would you put in the empty box so that...

a. there is no association between the variables?

b. children ages 9-11 are far more likely to walk to school than children ages 6-8?

c. children ages 6-8 are far more likely to walk to school than children ages 9-11?

	Walks to school?	
	yes	**no**
9-11 year olds	20	10
6-8 year olds	12	

41. **a.** Draw a line to fit the trend in the scatter plot below.

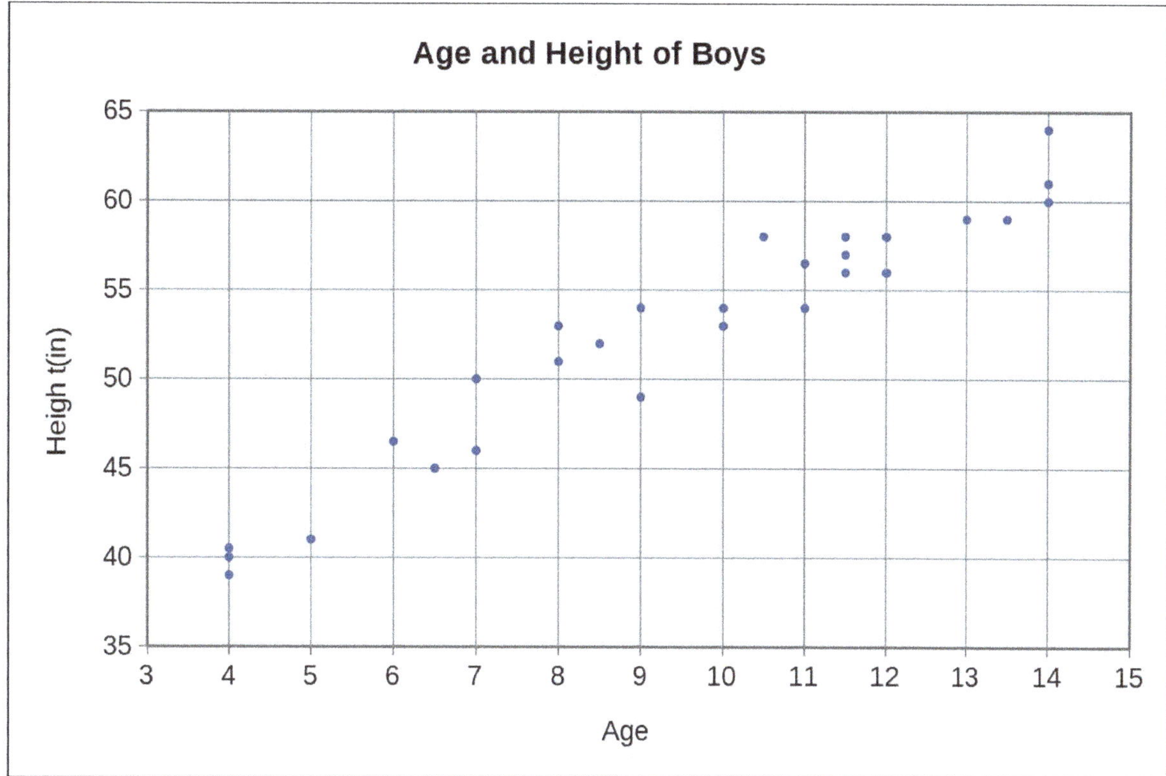

b. Find the (approximate) equation of ***your*** line.

c. A spreadsheet program calculates the equation for a trendline to be approximately $y = 2.1x + 32.6$. What does the slope of that equation signify in this context?

d. What does the y-interecept of that equation signify in this context?

e. What age does the equation from (c) predict for a boy that is 50.5 inches tall?

Answers

Exponents and Scientific Notation Review, p. 7

1.

a. $(-2)^4 = 16$	b. $-2^4 = -16$	c. $8^{-2} = 1/64$	d. $5^2 \cdot 5^8 \cdot 5^{-7} = 5^3 = 125$
e. $11 \cdot 10^{-2} = 11/100$	f. $10^3 + 10^4 = 11{,}000$	g. $\left(\frac{2}{-3}\right)^3 = -\frac{8}{27}$	h. $\frac{12^7}{12^5} = 12^2 = 144$

2.

a. $(a^{-1})^4 = \frac{1}{a^4}$	b. $(2x)^3 = 8x^3$	c. $(5x)^{-2} = \frac{1}{25x^2}$	d. $-2s^5t^7t^3 \cdot 4s^8 = -8s^{13}t^{10}$
e. $\frac{9a^7}{30a^5} = \frac{3a^2}{10}$ or $\frac{3}{10}a^2$	f. $\frac{x^3}{x^{-2}} = x^5$	g. $\left(\frac{3x}{-4}\right)^3 = -\frac{27x^3}{64}$	h. $\left(\frac{2a^2}{b}\right)^5 = \frac{32a^{10}}{b^5}$

3.

a. $8^x 8^5 = 8^{24}$ $x = 19$	b. $(7^8)^{-3} = \frac{1}{7^y}$ $y = 24$	c. $\left(\frac{2}{3}\right)^x = \frac{16}{81}$ $x = 4$	d. $\frac{(-3)^x}{(-3)^5} = -27$ $x = 8$	e. $(3^z)^2 = 9^4$ $z = 4$

4. (a) and (b) are true statements.

5. The error is on the next to the last line. It should say $2^2 \cdot 5^2 \cdot 2^3 \cdot 3^3$.
The final prime factorization then becomes $2^5 \cdot 3^3 \cdot 5^2$.

6. $x = 0$

7. a. 66 km/h (answer given to two significant digits)
b. 93 (answer rounded to the nearest meter, since the least accurate addend, 89 m, was accurate to the nearest meter.
c. \$14,300,000 or $1.43 \cdot 10^7$ dollars (answer given to three significant digits)

8. a. 3.5 km $\cdot$ 104 $\approx$ 360 km. The answer is given to two significant digits, since 3.5 has two, and 104 has three. If you think of 104 as coming from $52 \cdot 2$, the two is exact. It's not possible it could be 2.1 or 1.9. The two is actually exact, as in 2.000 or however many zeros you'd like to include.

b. 624.6 mi ÷ 24.3 gal ≈ 25.7 mi/gal c. 2.3 m $\cdot$ 11.9 m ≈ 27 m^2

9.

Planet	Average distance from sun (km)	In scientific notation (km)
Mercury	58,000,000	$5.8 \cdot 10^7$ km
Jupiter	778,570,000	$7.7857 \cdot 10^8$ km
Neptune	4,495,000,000	$4.495 \cdot 10^9$ km

10. a. $2.40 \cdot 10^{-4}$ seconds and $4.3 \cdot 10^{-3}$ seconds.
b. We subtract: $4.3 \cdot 10^{-3}$ s $-$ $2.40 \cdot 10^{-4}$ s = $43 \cdot 10^{-4}$ s $-$ $2.4 \cdot 10^{-4}$ s = $40.6 \cdot 10^{-4}$ s.
This needs rounded to $41 \cdot 10^{-4}$ s since 43 is accurate to the nearest one and 2.4 is accurate to the nearest tenth. Lastly, given in milliseconds, the answer is 4.1 milliseconds.

11. The cheetah is about $\frac{2.77 \cdot 10^1 \text{ m/s}}{1.3 \cdot 10^{-2} \text{ m/s}} \approx 2.1 \cdot 10^3 = 2{,}100$ times faster than the snail.

Exponents and Scientific Notation Test, p. 9

1. a. 81, −81   b. 5/8, 1,000   c. 25, 1/144   d. 1/64, 1/25

2.

a. $24v^8$	b. $-10b^8a^{15}$	c. $25x^2$	d. $\dfrac{1}{8x^3}$
e. $\dfrac{2}{s^2}$	f. $\dfrac{1}{s^6}$	g. $\dfrac{x^{10}}{32}$	h. $\dfrac{5x^5}{3}$

3. a, d, e

4. $0.4 \cdot 10^6$ $9 \cdot 10^5$ $5.2 \cdot 10^7$ $64 \cdot 10^6$ $2 \cdot 10^8$

5. a. $7 \cdot 10^9$ or 7,000,000,000 times bigger
   b. $2.5 \cdot 10^{10}$ or 25,000,000,000 times bigger

6. The area is 341 square feet (given to three significant digits, as the numbers in the problem).

7. $12{,}000 \cdot \$2.65 \cdot 5 \cdot 36 \approx \$5{,}720{,}000$. We give the answer to three significant digits because \$2.65 and 12,000 are accurate to three significant digits. The quantities 5 days and 36 weeks, being perfectly exact, are accurate to any number of significant digits, so they do not affect how we round the final answer.

8. 1.5 km/s · 60 s/min · 60 min/hr · 24 hr/day · 365 day/year = 47,304,000 km/year ≈ $4.7 \cdot 10^7$ km/year
   Sounds travels $4.7 \cdot 10^7$ km in one year, in salt water.

9. We divide 10 grams by the mass of one copper atom to find the number of atoms in 10 grams of copper:

$$\frac{10 \text{ g}}{1.055 \cdot 10^{-22} \text{ g}} = 0.948 \cdot 10^{23} \text{ atoms} = 9.48 \cdot 10^{22} \text{ atoms}$$

Geometry Review, p. 11

1. Answers will vary. Check the student's answer.
   For example: First rotate the figure around the origin 180 degrees. Then translate it one unit to the right.

2.

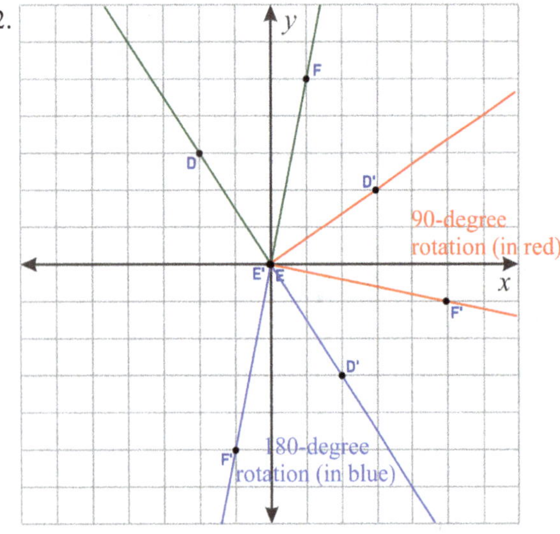

3. The coordinates of its vertices before these transformations were (−5, 3), (−2, 5), (−1, 4), and (−4, 1).

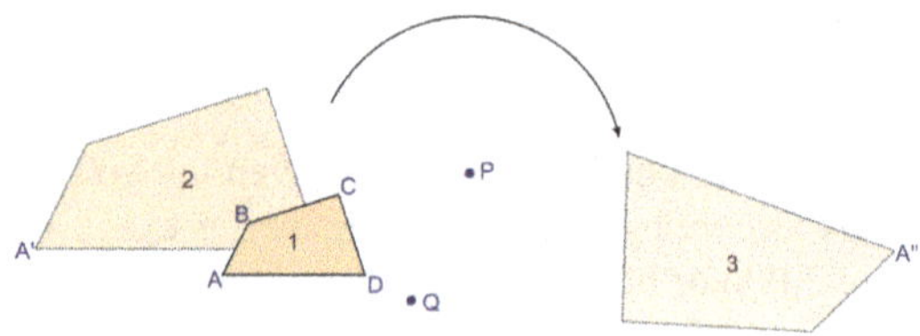

4. The true statements are (a), (c), and (d).

Geometry Review, cont.

5. a.

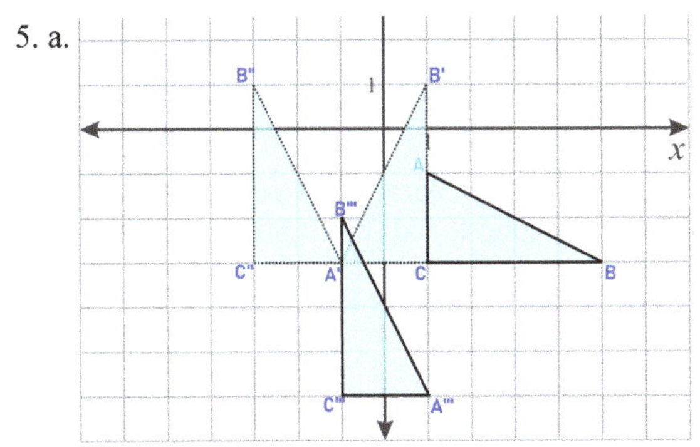

b. Answers will vary. Check the student's answer. For example: First, reflect the triangle in the horizontal line $y = -3$. Then rotate it 90° counter-clockwise around point C'. Lastly translate it three units down and two units to the left.

6. a.

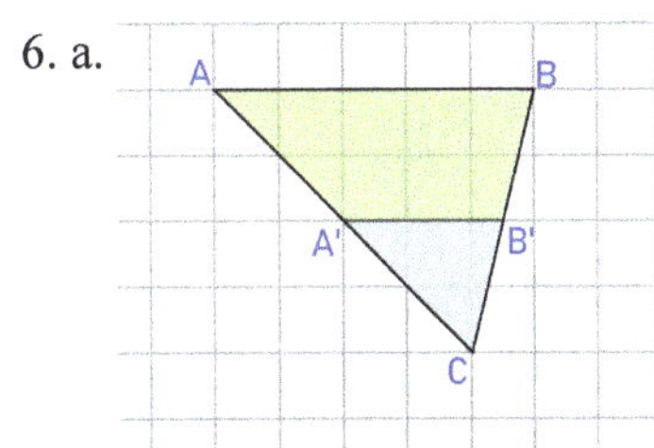

b.

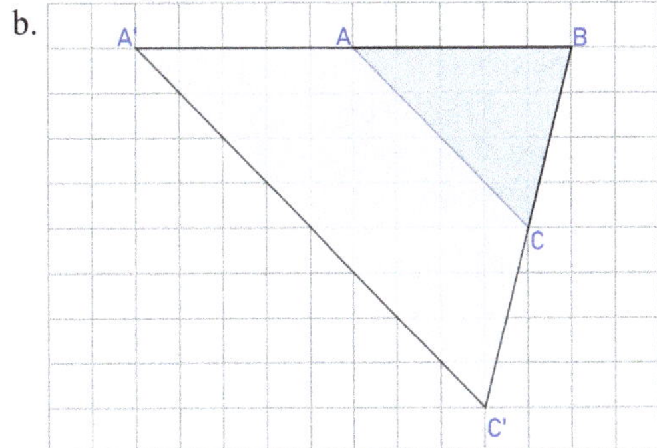

7. The coordinates of point F''' are (−4, −1).

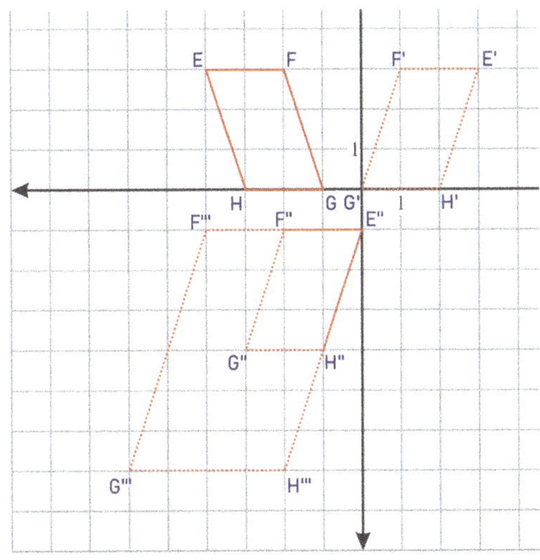

8. Answers will vary. Check the student's answer. For example: First, rotate the triangle 90 degrees clockwise around point E. Then dilate it with scale factor 1/3 from point E. Lastly, translate it four units to the right.

9. The dilation is from point S, since S' is the same as S. To figure out the scale factor, look at the distances from P to S, and then from P' to S'. From P to S is one unit horizontally and two units vertically. From P' to S' is two units horizontally and four units vertically, so the distances doubled. Thus, the scale factor is 2.

Since S' = S'', the rotation is around point S' (or S). And since R' is six units horizontally from S', and R'' is six units vertically from S'', the rotation is 90° clockwise around S'.

Original figure	Dilation	Rotation
P(−5, 3)	P'(−6, 5)	P''(0 , 3)
Q(0, 3)	Q'(4, 5)	Q''(0 , −7)
R(−1, 1)	R'(2 , 1)	R''(−4 , −5)
S(−4, 1)	S'(−4, 1)	S''(−4, 1)

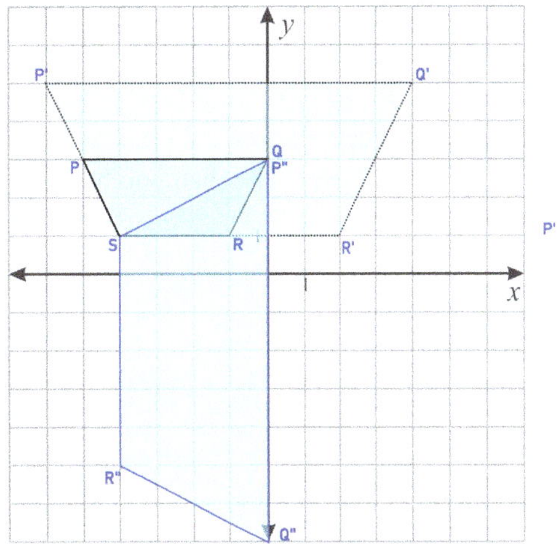

10. a. The three angles of the triangle are $x + 13$, $x + 18$, and $5x + 2$ (note there are vertical angles formed; that is why the unmarked angle of the triangle is $x + 13$). Those add up to 180°, so we can write the equation:

$$\begin{aligned} x + 13 + x + 18 + 5x + 2 &= 180 \\ 7x + 33 &= 180 \\ 7x &= 147 \\ x &= 21 \end{aligned}$$

b. Angle α is supplementary to the angle marked with $x + 13$. Since x is 21, $x + 13$ is 34°. So, α = 180° − 34° = 146°.

11. In the larger triangle, the third angle, angle α', is 180° − 85° − 57° = 38°.

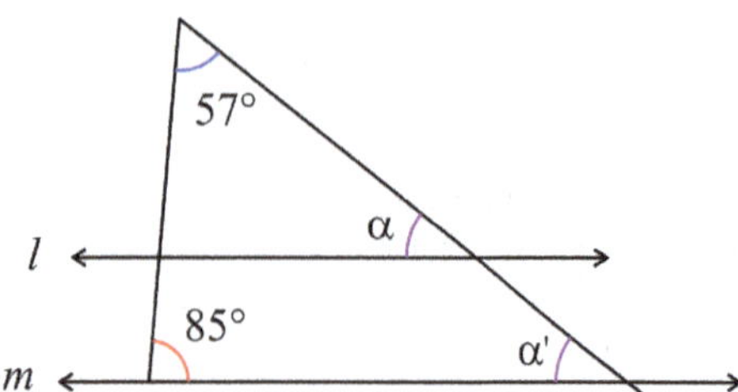

This is also the measure of the angle α, since α and α' are corresponding angles. So, α = 38°.

12. Since m and n are parallel lines, angle A and the 64° angle are corresponding angles, thus congruent.

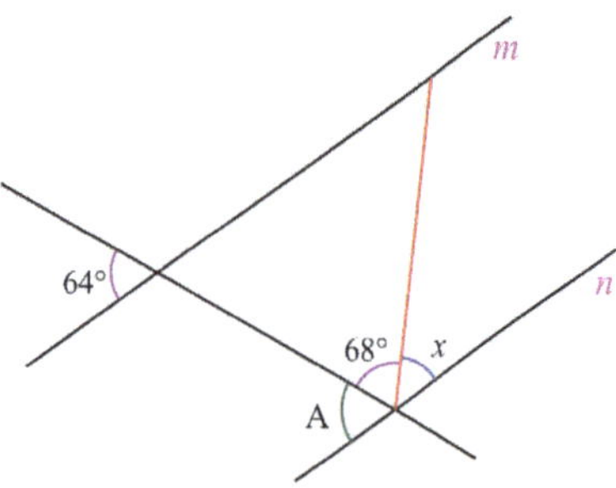

Angle A, the 68° angle, and x form a straight line so their sum is 180°. From that, we can get that x = 180° − 68° − 64° = 48°.

13. Its volume in cubic centimeters is $\pi \cdot (3\text{ cm})^2 \cdot 17$ cm $\approx 480.66\text{ cm}^3$ = 480.66 ml. The shampoo bottle is about 473/480.66 ≈ 98% full.

14. The volume of the sphere is $(4/3) \cdot \pi \cdot (5\text{ cm})^3$. The volume of a cone is $A_b \cdot h/3$, where A_b is the area of the base. In this case, A_b equals $\pi \cdot (5\text{ cm})^2$.

Now we simply set the volume of the cone equal to the volume of the sphere, and solve for the height (h).

$$A_b \cdot h / 3 = (4/3) \cdot \pi \cdot (5\text{ cm})^3$$
$$\pi \cdot (5\text{ cm})^2 \cdot h / 3 = (4/3) \cdot \pi \cdot (5\text{ cm})^3$$

Now, π cancels out from both sides:

$$(5\text{ cm})^2 \cdot h / 3 = (4/3) \cdot (5\text{ cm})^3$$
$$25\text{ cm}^2 \cdot h / 3 = (4/3) \cdot 125\text{ cm}^3$$

Dividing both sides by 25 cm^2, we get

$$h / 3 = (4/3) \cdot 5\text{ cm}$$
$$h = 4 \cdot 5\text{ cm} = 20\text{ cm}$$

So, when the height is 20 cm, the cone will have an equal volume to the sphere with a radius of 5 cm.

Another way to solve this is to actually calculate the volume of the sphere and the area of the base of the cone, and use those calculated numbers in the equation.

The volume of the sphere is $(4/3) \cdot \pi \cdot (5\text{ cm})^3 \approx 523.599\text{ cm}^3$. The area of the base of the cone is $\pi \cdot (5\text{ cm})^2 \approx 78.5398\text{ cm}^2$.

Then, our equation looks like this:

$$A_b \cdot h / 3 = (4/3) \cdot \pi \cdot (5\text{ cm})^3$$
$$78.5398\text{ cm}^2 \cdot h / 3 = 523.599\text{ cm}^3$$
$$78.5398\text{ cm}^2 \cdot h = 3 \cdot 523.599\text{ cm}^3$$
$$h = \frac{3 \cdot 523.599\text{ cm}^3}{78.5398\text{ cm}^2}$$
$$h \approx 20.0\text{ cm}$$

15. The mound of ammonium nitrate consists of two parts: a cylinder and a cone.
The volume of the cylinder is $\pi \cdot (57.5\text{ ft})^2 \cdot 36$ ft.
The volume of the cone is $\pi \cdot (57.5\text{ ft})^2 \cdot 57.5\text{ ft} / 3$.

The total volume is
$\pi \cdot (57.5\text{ ft})^2 \cdot 36\text{ ft} + \pi \cdot (57.5\text{ ft})^2 \cdot 57.5\text{ ft} / 3$

≈ 573,000 cubic feet.

Geometry Test, p. 16

1.

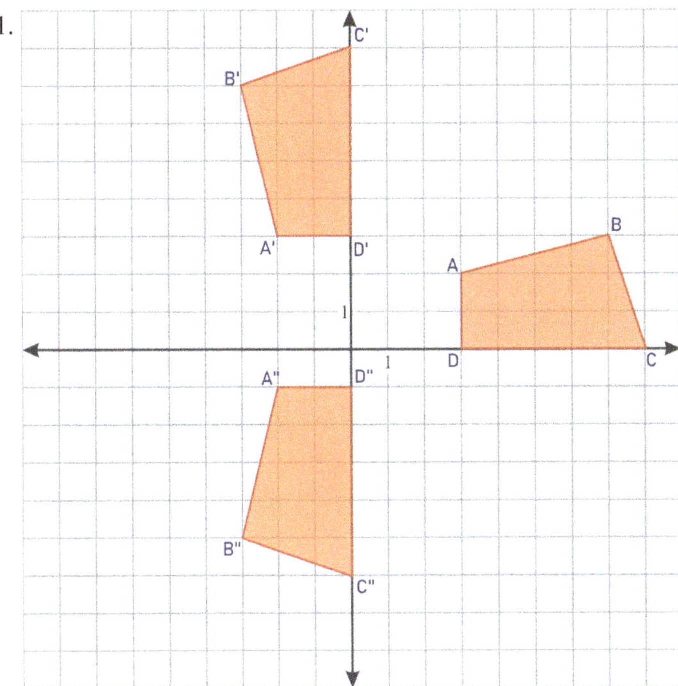

2. Student answers may vary; check the student's answer. At least three transformations are necessary: a dilation with a scale factor of 1/2 (can be from a variety of points), a rotation, and a translation. For example:

   First, dilate the bigger arrow from point A and with scale factor 1/2. Then, rotate the resulting image 90 degrees clockwise around the origin. Lastly, translate the figure four units down.

   Or, first rotate the large arrow 90 degrees clockwise around the origin. Then dilate it from point A′ and with scale factor 1/2. Lastly translate the figure four units down.

3. (4, 0), (3, 2), and (1, 0)

4. The points are (2, −2), (2, −1), (0, −2) and (0, −1).

5. a. rotation
   b. and c.

   d. Only (iv) and (v) are preserved.

6. Student reasoning may vary; please check the student's reasoning. For example:

   Angle BCD is supplementary to the 115° angle, so it is 65°. Angle CBD is 42° because it is an alternate interior angle with the given 42° angle. Angle x is the third angle in the triangle BCD, so it is 180° − 65° − 42° = <u>73°</u>.

7. Angle BAC is supplementary to the 150° so it measures 30°. Angle ABC is the third angle in triangle ABC, so it is 180° − 30° − 77° = 73°. Lastly, x is a vertical angle with angle ABC, so, it also measures <u>73°</u>.

Geometry Test, cont.

8. Yes, they are. The third angle in the smaller triangle measures 180° – 25° – 60° = 95°. And the third angle in the larger triangle measures 180° – 25° – 95° = 60°. So, both triangles have angles that measure 25°, 60°, and 95°. This means they are similar triangles.

9. a. $V = (4/3)\pi\ (4.0\text{ cm})^3 \approx 270\text{ cm}^3$

   b. $V = \pi\ (4.0\text{ cm})^2 \cdot 16\text{ cm} \approx 800\text{ cm}^3$.

10. The volume of the cone is <u>one-third</u> of the volume of the cylinder. You don't need to calculate the volumes to find that out, because the formulas for the volumes show this fact easily. The volume of a cylinder is $A_b h$ where A_b is the area of the bottom face and h is its height. The volume of a cone is $(1/3)A_b h$ or 1/3 of the volume of the cylinder with the same bottom face and the same height.

11. Yes, you should believe her. The volume of one lime is $(4/3)\pi\ (3\text{ cm})^3 \approx 113\text{ cm}^3$ or 113 ml. The volume of the four limes is 452 ml, which is more than 1 1/2 cups. Now, the lime is not 100% juice; it also has rind and other material inside the fruit. Still, it is reasonable to expect to get half a cup of juice from these four limes. (On average, four limes do give you half a cup of lime juice.)

Mixed Review 1, p. 20

1. a. –16 b. 1/125 c. 57/1000 d. 3
   e. 1/64 f. 1/25 g. $5 \cdot 2^3 + 4 = 44$

2. a. 10,000 b. 2/343 c. 1

3.

a. $8b^{15}$	b. $\dfrac{1}{27y^3}$	c. $\dfrac{6}{x^2}$	d. $-10a^{10}b^9$
e. $\dfrac{1}{x}$	f. $\dfrac{2}{a^3}$	g. $\dfrac{-8x^3}{27}$	h. $\dfrac{x^2}{z^4}$

4. a. > b. < c. <

5. No. The quantity $5x$ is in parentheses, and this means even the number 5 is raised to the power of –2. Compare:

$$(5x)^{-2} = \frac{1}{(5x)^2} = \frac{1}{25x^2} \neq 25x^{-2}.$$

6. a. 3 b. 2 c. 2 d. 3
   e. 1 f. 4 g. 3 h. 4

7. a. $4.03 \cdot 10^{-1}$ b. $6.6 \cdot 10^7$
   c. $2 \cdot 10^{-3}$ d. $2.91 \cdot 10^{-2}$

8.

a. $2 \cdot 10^{-3} + 5 \cdot 10^{-2}$ $= 0.002 + 0.05 = \underline{0.052}$	b. $7 \cdot 10^{-5} + 0.03$ $= 0.00007 + 0.03 = \underline{0.03007}$
c. $3.2 \cdot 10^{-1} - 0.07$ $= 0.32 - 0.07 = \underline{0.25}$	d. $5.4 \cdot 10^4 - 2{,}000 + 8 \cdot 10^3$ $= 54{,}000 - 2{,}000 + 8{,}000$ $= \underline{60{,}000}$

9. a. 1.2 m^2 b. 25 mi/gal
   c. 7,930 mi d. 58,000 lb
   e. \$350/person f. 72,000 kg

10. $$\frac{6.09 \cdot 10^{12}\text{ km}^2}{12 \cdot 10^3} = 5.1 \cdot 10^8\text{ km}^2$$

11. a. Since 550,000 is accurate to the thousands, it has three significant digits, and our answer also needs to be given with three significant digits. They spent \$1,780 per student.

    b. 950 cm^2. Notice the answer needs to be given to two significant digits.

12. $$\frac{9.3 \cdot 10^7\text{ mi}}{1.38 \cdot 10^{-1}\text{ hr}} = 6.7 \cdot 10^8\text{ mph}$$

    — which is the speed of light! The light of the sun reaches the earth in about 0.138 hours which is about 8 minutes and 17 seconds.

Mixed Review 2, p. 22

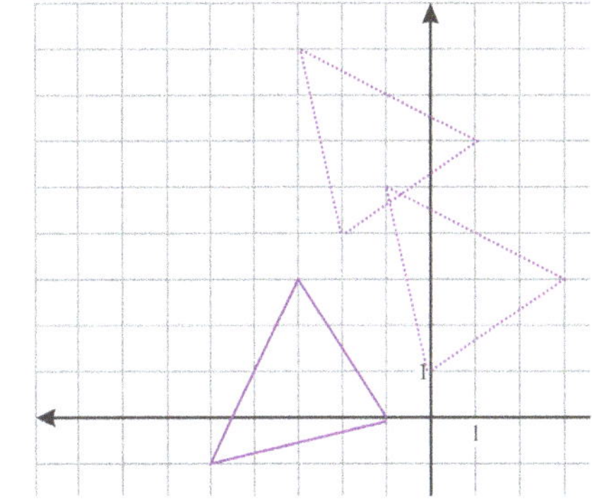

1. (−3, 8), (1, 6), and (−2, 4). See the image on the right.

2. (14, 9), (14, 13), (11, 9), and (11, 13)

   We start with (12, −6), (12, −10), (9, −6), and (9, −10), and do the transformations in the backwards order. First, after the reflection, the coordinates are (12, 6), (12, 10), (9, 6), and (9, 10).

   Then we move it 3 units up and 2 to the right, and get (14, 9), (14, 13), (11, 9), and (11, 13).

3.

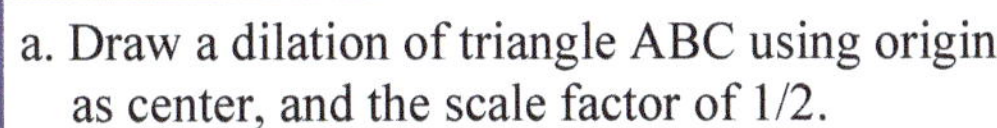

a. Draw a dilation of triangle ABC using origin as center, and the scale factor of 1/2.

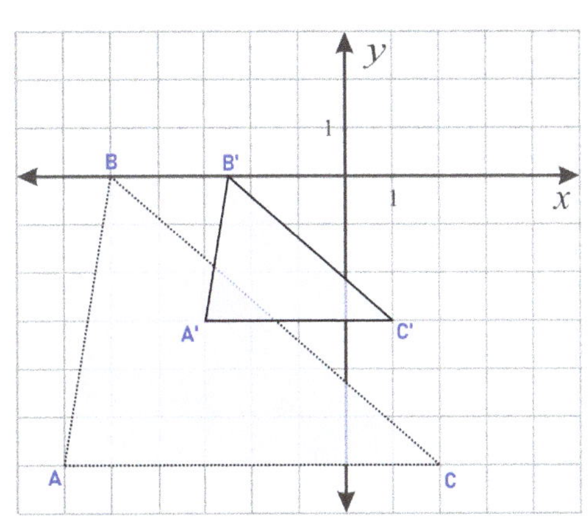

b. Draw a dilation of triangle ABC from point B, again using the scale factor of 1/2.

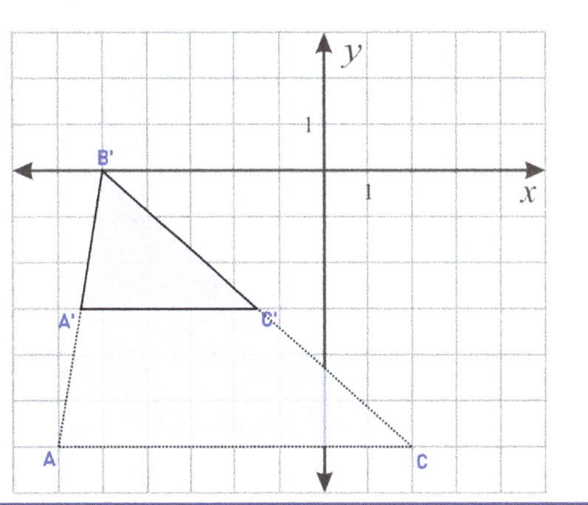

4. a. −16 b. 1/36 c. 3/1000 d. 1/5 e. 1/27,000 f. 117 g. 25/36 h. 1/121

5.

a. $\frac{1}{5s^2}$	d. $\frac{1}{x^5y^{15}}$	g. $8a$
b. $\frac{t^6}{16}$	e. $\frac{1}{9x^2y^6}$	h. $\frac{16y^2}{-1000y^3} = -\frac{2}{125y}$
c. $\frac{1}{x^2y^6}$	f. $\frac{16}{w^8}$	i. $\frac{27x^3}{-3x} = -9x^2$

6. a. $x = 4$ b. $y = 3$ c. $x = 4$

7.

$0.07 \cdot 10^9$	$7 \cdot 10^4$	$7 \cdot 10^9$
70,000,000	$70 \cdot 10^8$	$0.7 \cdot 10^5$
$0.7 \cdot 10^8$	7,000,000,000	70,000

8. a. $9.3 \cdot 10^6$ b. $2.3 \cdot 10^7$ c. $1.45 \cdot 10^9$

9. a. It means $2.1 \cdot 10^{25}$.
   b. Answers will vary; check the student's answer. For example: $7 \cdot 10^{12}$ and $3 \cdot 10^{12}$ or $7 \cdot 10^6$ and $3 \cdot 10^{18}$.

10. The population of Lydia's hometown is $\frac{1.5 \cdot 10^9}{2 \cdot 10^4} = 0.75 \cdot 10^5 = 75{,}000$.

11. a. 20 m^2 b. 120 kg
    c. \$24.20/hr d. 36.4 mi/gal

Linear Equations Review, p. 24

1.

a.
$$\begin{aligned} 7x - 3(x-5) - 2x &= 10 \\ 7x - 3x + 15 - 2x &= 10 \\ 2x + 15 &= 10 && \mid -15 \\ 2x &= -5 && \mid \div 2 \\ x &= -2\tfrac{1}{2} \end{aligned}$$

b.
$$\begin{aligned} x - \frac{1}{3} &= \frac{3}{4}x - 2 && \mid \cdot 12 \\ 12x - 4 &= 9x - 24 && \mid -9x \\ 3x - 4 &= -24 && \mid +4 \\ 3x &= -20 && \mid \div 3 \\ x &= -6\tfrac{2}{3} \end{aligned}$$

c.
$$\begin{aligned} 20 - q &= -q + 2(q-5) - 6q \\ 20 - q &= -q + 2q - 10 - 6q \\ 20 - q &= -10 - 5q && \mid +5q \\ 20 + 4q &= -10 && \mid -20 \\ 4q &= -30 && \mid \div 4 \\ q &= -7\tfrac{1}{2} \end{aligned}$$

d.
$$\begin{aligned} -52 - 2(x+14) &= 80 - 11x + x \\ -52 - 2x - 28 &= 80 - 10x \\ -80 - 2x &= 80 - 10x && \mid +10x \\ -80 + 8x &= 80 && \mid +80 \\ 8x &= 160 && \mid \div 8 \\ x &= 20 \end{aligned}$$

2. Let x be Heather's age now. The equation is: $x - 5 = (1/3)(x + 45)$, or it can also be written as $x - 5 = (x + 45)/3$.

See the solution on the right.

$$\begin{aligned} x - 5 &= \frac{x+45}{3} && \mid \cdot 3 \\ 3x - 15 &= x + 45 && \mid -x \\ 2x - 15 &= 45 && \mid +15 \\ 2x &= 60 && \mid \div 2 \\ x &= 30 \end{aligned}$$

3.

a.
$$\begin{aligned} \frac{x-1}{2} &= \frac{3x+2}{7} && \mid \text{cross-multiply} \\ 7x - 7 &= 6x + 4 && \mid -6x \\ x - 7 &= 4 && \mid +7 \\ x &= 11 \end{aligned}$$

b.
$$\begin{aligned} x - \frac{1-3x}{2} &= 5 && \mid \cdot 2 \\ 2x - (1 - 3x) &= 10 \\ 2x - 1 + 3x &= 10 \\ 5x - 1 &= 10 && \mid +1 \\ 5x &= 11 && \mid \div 5 \\ x &= 11/5 = 2\tfrac{1}{5} \end{aligned}$$

4. a.
$$\begin{aligned} \rho &= \frac{m}{V} && \mid \cdot V \\ V\rho &= m && \mid \div \rho \\ V &= \frac{m}{\rho} \end{aligned}$$

b. $V = m/\rho = 4.6\text{ kg}/(850\text{ kg/m}^3) \approx 0.0054\text{ m}^3 = 5{,}400\text{ cm}^3$.

The conversion factor between cubic meters and cubic centimeters:
$1\text{ m}^3 = 1\text{ m} \times 1\text{ m} \times 1\text{ m} = 100\text{ cm} \times \underline{100}\text{ cm} \times \underline{100}\text{ cm} = \underline{1{,}000{,}000}\text{ cm}^3$

5. Answers will vary; check the student's answers. For example:

a. $8x - 2 = 5 + 8x$ OR $8x - 2 = 4(2x + 6)$
b. $8x - 2 = 6x + 7$ OR $8x - 2 = 3(3x - 1)$
c. $8x - 2 = 3x + 5x + 8 - 10$ OR $8x - 2 = 2(4x - 1)$

Linear Equations Review, cont.

6.

a.			b.			
$0.6 + 9.4x - 2 = x - 4.8x$			$4 - 0.3(2y + 2.8) = 1.5(y - 3)$	
$9.4x - 1.4 = -3.8x$	$+ 3.8x$		$4 - 0.6y - 0.84 = 1.5y - 4.5$	
$14.2x - 1.4 = 0$	$+ 1.4$		$3.16 - 0.6y = 1.5y - 4.5$	$+ 0.6y$
$14.2x = 1.4$	$\div 14.2$		$3.16 = 2.1y - 4.5$	$+ 4.5$
$x \approx 0.10$			$7.66 = 2.1y$	$\div 2.1$
			$y \approx 3.65$	

7. Let n be the number of cheaper cartons of almond milk. Then, $16 - n$ is the number of the more expensive ones.

$$\begin{aligned} 4.5n + 5.2(16 - n) &= 76.2 \\ 4.5n + 83.2 - 5.2n &= 76.2 \\ 83.2 - 0.7n &= 76.2 && \big| - 83.2 \\ -0.7n &= -7 && \big| \div (-0.7) \\ n &= 10 \end{aligned}$$

8. Let x be the increase in price.

$$\begin{aligned} 0.75(8.5 + x) &= 7.5 \\ 6.375 + 0.75x &= 7.5 && \big| - 6.375 \\ 0.75x &= 1.125 && \big| \div 0.75 \\ x &= 1.5 \end{aligned}$$

You should increase the price by \$1.50.

9. Let p be the original price.

$$\begin{aligned} 0.65 \cdot 1.07p &= 53.90 \\ 0.6955p &= 53.9 && \big| \div 0.6955 \\ x &\approx 77.4982 \end{aligned}$$

The original price was \$77.50.

Linear Equations Test, p. 27

1.

a.	b.
$6 - 3y - 2 + 8y = 4y + 1 - 9y - 5$	$12 - (x - 2) = 10 - 2x$
$5y + 4 = -5y - 4$	$12 - x + 2 = 10 - 2x$
$10y + 4 = -4$	$14 - x = 10 - 2x$
$10y = -8$	$14 + x = 10$
$y = -4/5$	$x = -4$

c.	d.
$4x + 20 - x = 2(x - 5) - 6x$	$-35 - 3(x - 4) = 10x + 40 - x + 13x$
$4x + 20 - x = 2x - 10 - 6x$	$-35 - 3x + 12 = 10x + 40 - x + 13x$
$3x + 20 = -4x - 10$	$-3x - 23 = 22x + 40$
$7x + 20 = -10$	$-25x - 23 = 40$
$7x = -30$	$-25x = 63$
$x = -30/7 = -4\ 2/7$	$x = -63/25 = -2\ 13/25$

Linear Equations Test, cont.

2. R = V/C

3. Let L be Lucas's age now. The equation is L + 6 = (2/3)(66 + 6), or L + 6 = (2/3) 72, from which L = 42.

4. Let x be the increase in price. One way to write an equation from this situation is $0.8(288 + x) = 249$, from which $230.4 + 0.8x = 249$, and $x = 23.25$. He should increase the price by $23.25.

5. a. It has one solution.
   b. Answers will vary; check the student's answer. For example, the equation could be changed to $-8y + 12 = 4(3 - 2y)$ or to $8y + 4 = 4(1 + 2y)$, just anything where everything cancels out, so that the equation reduces to $0 = 0$.

6. a. It has an infinite number of solutions.
   b. Answers will vary; check the student's answer. For example, the equation could be changed to $-2(9y - 3) + 8y = 8 - 10y$ or $-2(9y - 4) + 8y = 2 - 10y$. The terms with y need to cancel out, but the constant terms should not.

7.

a.	b.
$\frac{3x-5}{4} - 1 = 2x$	$\frac{y-2}{3} = 3y + \frac{5-y}{2}$
$3x - 5 - 4 = 8x$	$2y - 4 = 18y + 15 - 3y$
$3x - 9 = 8x$	$2y - 4 = 15y + 15$
$-5x - 9 = 0$	$-13y - 4 = 15$
$-5x = 9$	$-13y = 19$
$x = -9/5 = -1\ 4/5$	$y = -19/13 = -1\ 6/13$

Mixed Review 3, p. 29

1. a. $8^6 = (2^3)^6 = 2^{18}$	b. $64^3 = (4^3)^3 = 4^9$	c. $100^5 = (10^2)^5 = 10^{10}$

2.

a.	b.	c.	d.
$5 \cdot 4^3 = 320$	$(-5)^3 = -125$	$-3^{-2} = -1/9$	$2 \cdot 5^{-2} = 2/25$
$(5 \cdot 4)^3 = 8{,}000$	$-5^3 = -125$	$(-3)^{-2} = 1/9$	$(2 \cdot 5)^{-2} = 1/100$

3. About $3 \cdot 10^5$ times the mass of the earth. (The calculation gives us $(1/3) \cdot 10^6$, and to give this in proper scientific notation, we write it as $3 \cdot 10^5$, with one significant digit, just like the numbers in the problem.)

4.

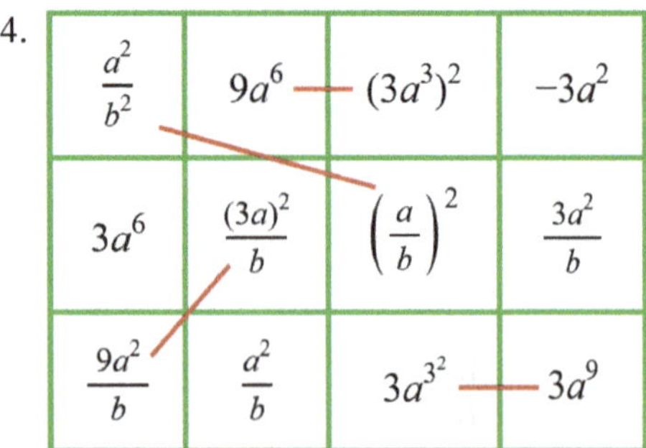

$\frac{a^2}{b^2}$	$9a^6$	$(3a^3)^2$	$-3a^2$
$3a^6$	$\frac{(3a)^2}{b}$	$\left(\frac{a}{b}\right)^2$	$\frac{3a^2}{b}$
$\frac{9a^2}{b}$	$\frac{a^2}{b}$	$3a^{3^2}$	$3a^9$

5. The third angle of the triangle is 180° − 39° − 74° = 67°. The angle x is exterior to the 67° angle, so it measures 180° − 67° = 113°.

Mixed Review 3, cont.

6. The area of the bottom triangle is 10 cm · 15 cm ÷ 2 = 75 cm^2. The volume is 36 cm · 75 cm^2 = 2,700 cm^3. (We do not need the fact that the longest side of the triangle is 18 cm.)

7. In the image, angle 1 and angle x are supplementary, and therefore $\angle 1 = 180° - 76° = 104°$. Then, angle 1 and y are alternate interior angles, thus congruent. So, $y = 104°$.

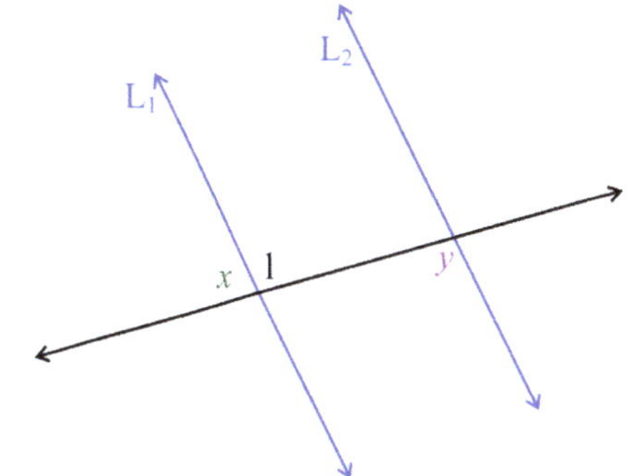

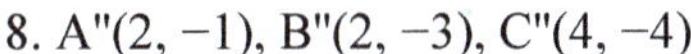

8. A"(2, −1), B"(2, −3), C"(4, −4)

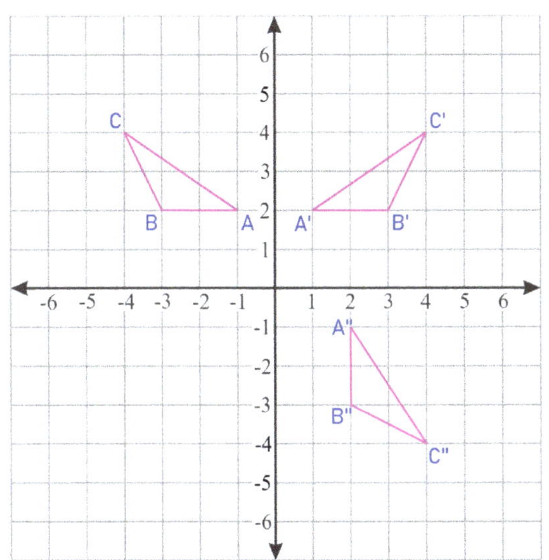

9.

a. scale factor 3

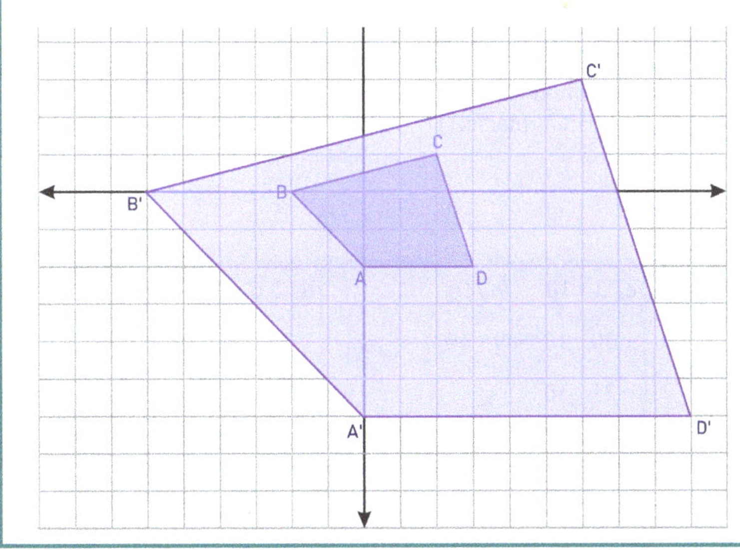

b. scale factor 1/2

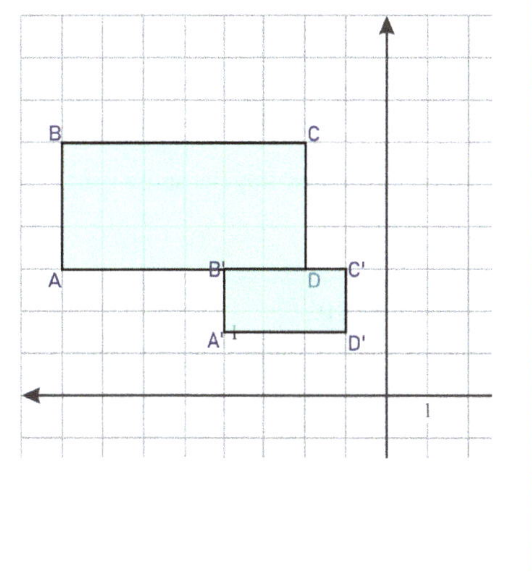

Mixed Review 4, p. 31

1. Answers will vary; check the student's answer. For example:

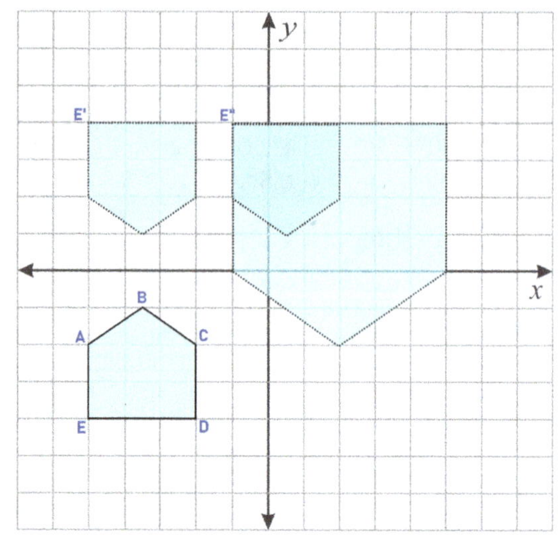

First, reflect the pentagon in the x-axis. Then, translate it four units to the right. Lastly, dilate it from point E" and with a scale factor of 2. (See the image on the right.)

Or, first, rotate the figure 180 degrees around the origin. Then dilate it from the point E' with scale factor 2.

There are many other possible ways to do it. Each sequence should include a reflection in the x-axis or a rotation 180 degrees around the origin, a dilation with a scale factor of 2, and a possible translation. These can be in different orders.

2.

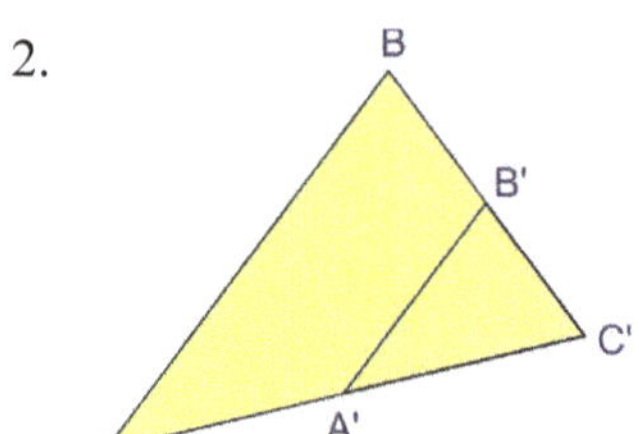

3. Please refer to the image on the right. The student will also need to name angles, points, and/or lines in the figure to be able to reference them in the explanation.

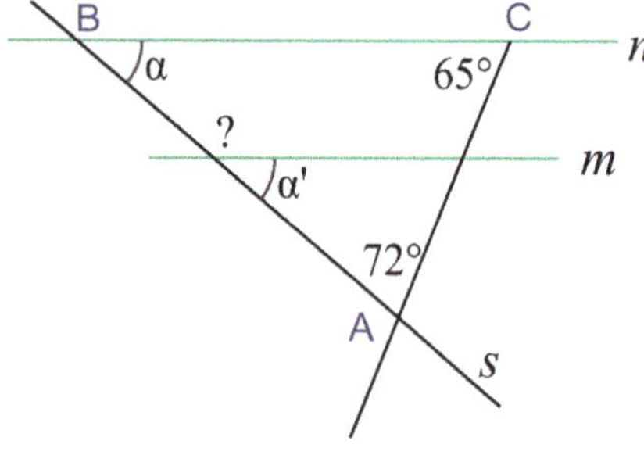

Student explanations will vary; please check the student's explanation.

For example:
Points A, B, and C mark the triangle ABC. In triangle ABC, $\angle\alpha = 180° - 72° - 65° = 43°$. Since lines n and m are parallel and line s is a transversal to them, α and α' are corresponding angles, and thus congruent. The angle in question is supplementary to α, so it is 180° − 43° = <u>137°</u>.

4. a. The first one is a rotation. The second is a reflection. (The reflection is not in the x or y-axis. It is a reflection in a line — a line that goes through the points where $\overline{B'C'}$ and $\overline{B''C''}$ intersect).

b. All the attributes stay the same except position.

5.

a. $2 \cdot 10^8 + 8 \cdot 10^7$	b. $1.3 \cdot 10^6 + 2 \cdot 10^7$	c. $3 \cdot 10^5 - 9 \cdot 10^4$	d. $2.6 \cdot 10^8 - 5 \cdot 10^7$
$= 20 \cdot 10^7 + 8 \cdot 10^7$	$= 1.3 \cdot 10^6 + 20 \cdot 10^6$	$= 30 \cdot 10^4 - 9 \cdot 10^4$	$= 26 \cdot 10^7 - 5 \cdot 10^7$
$= 28 \cdot 10^7$	$= 21.3 \cdot 10^6$	$= 21 \cdot 10^4$	$= 21 \cdot 10^7$
$= 2.8 \cdot 10^8$	$= 2.13 \cdot 10^7$	$= 2.1 \cdot 10^5$	$= 2.1 \cdot 10^8$

6. (b) is in error. It should be $(2 \cdot 10)^{-3} = \dfrac{1}{8 \cdot 10^3} = \dfrac{1}{8000}$.

7. Let h be the height of the cone. The volume of the cone is $(1/3) \cdot \pi \cdot (6 \text{ ft})^2 \cdot h$, and on the other hand, the volume is 11.0 cubic yards, or $11 \cdot 27 = 297$ cubic feet. So, we can write an equation:

$$\frac{\pi \cdot (6 \text{ ft})^2 h}{3} = 297 \text{ ft}^3$$

$$\pi \cdot 36 \text{ ft}^2 \cdot h = 3 \cdot 297 \text{ ft}^3$$

$$h = \frac{3 \cdot 297 \text{ ft}^3}{\pi \cdot 36 \text{ ft}^2}$$

$$h = 7.88 \text{ ft}$$

Mixed Review 4, cont.

8. a. 14.8 km (to the accuracy of the least accurate measurement, which was to the tenth of a kilometer)

   b. 448 mi (to the accuracy of the least accurate measurement, which was to the mile)

9. 170 meters (to two significant digits, just as 4.2 cm has two significant digits)

10. $2.2 \cdot 10^9$ times

11. The time it takes is $t = \dfrac{d}{v} = \dfrac{4.0208 \cdot 10^{13} \text{ km}}{3.00 \cdot 10^5 \text{ km/s}} \approx 1.340267 \cdot 10^8$ seconds. To convert this to years, we need to divide it by $60 \cdot 60 \cdot 24 \cdot 365$. Doing that, we get $\dfrac{1.340267 \cdot 10^8 \text{ seconds}}{60 \cdot 60 \cdot 24 \cdot 365} = \underline{4.25 \text{ years}}$.

    Note that this is given to three significant digits, since the speed of light was given to three significant digits (and the distance was given to five.)

Introduction to Functions Review, p. 34

1. Answers will vary; check the student's answer. The problems are: One needs to add an age for Max, and Luna should not have two ages (one needs removed). For example:

Input	Output
Name	**Age**
Fifi	2
Bella	5
Max	3
Luna	2
Charlie	6

2. Because for the input 2, there are two outputs (10 and 20).

3. In the x-interval $[-10, -5]$, the function is increasing in a nonlinear manner.
   In the interval $[-5, 1]$, the function is decreasing in a nonlinear manner.
   In the interval $[1, 4]$, the function is increasing in a linear manner.
   In the interval $[4, 9]$, the function is constant.

4. Answers will vary; check the student's answer. For example:

Time (hours)	0	1	2	3	4	5	6	7	8	9	10	11	12
Cost ($)	0	5	6	7	10	12	14	17	18	18	20	20	20

5. a. No, it is not linear, because the graph is not a line.
   b. $5,000/yr
   c. $2,000/yr

6. a. The rates of change are as follows. Function 1: 0.22 miles/minute.
   Function 2: (0.45 mi)/(2 min) = 0.225 miles/minute. Function 2 has the greater rate of change.
   This means Jayden's sister was bicycling with a faster speed during that time.
   b. Function 1 is linear, and function 2 is nonlinear.
   c. Jayden takes (1.5 mi)/(0.22 mi/min) $\approx$ 6.82 minutes to reach home. His sister is bicycling at a rate of 0.12 miles/minute between 5 and 6 minutes, which means that at 7 minutes, she will have covered 1.36 miles, and at 8 minutes, 1.48 miles. But Jayden has already reached home by 7 minutes of time, so he reached home first.

Introduction to Functions Review, cont.

7. a. 3 inches/hour. It means the snow is falling at that rate.
   b. 17 inches. Before the blizzard started, there was already 17 inches of snow on the ground.
   c. $S = 17 + 3t$ where S is the amount of snow, and t is time in hours
   d. From the equation $26.5 = 17 + 3t$, we can solve that $t = 3\ 1/6$ hours, or 3 hours 10 minutes.
   At 5:40, the snow was 26.5 inches deep.
   e. 9 PM is 6.5 hours after the blizzard started, so the depth of snow is $S = 17 + 3(6.5) = 36.5$ inches deep.
   f. Answers will vary; check the student's graph. For example:

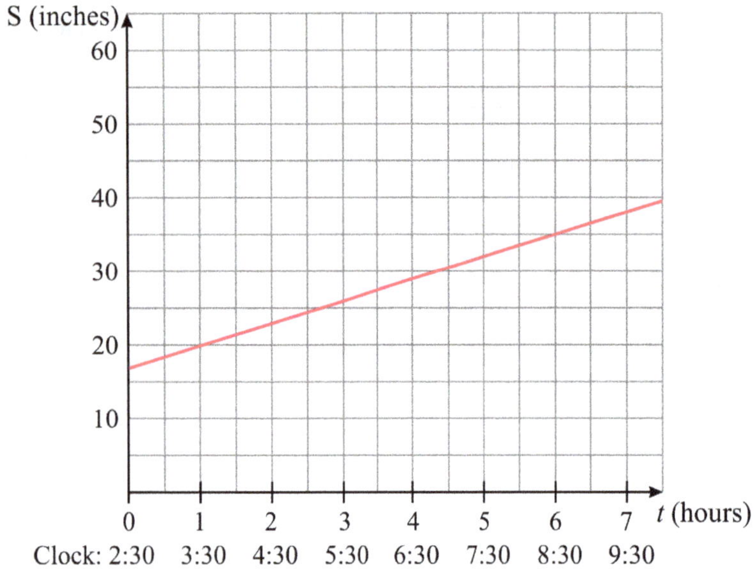

8. Answers will vary; check the student's story. For example: Maria sits at 12 feet from the blackboard. Two minutes into the class, the teacher calls her to the board to solve a math problem, so Maria quickly walks there. She solves the problem, and after two minutes, goes back to her desk. Then, 6 minutes into the class, she goes to the back of the room to sharpen her pencil, which takes less than a minute. She then returns to her desk and stays seated.

9. a. Greg.
   b. They meet from 20 to 25 minutes, and also at about 32 minutes.
   Between 20 and 25 minutes, they have traveled 1.5 miles. At 32 minutes, they have traveled about 1.9 miles.
   c. Trevor finishes first, about 2 minutes quicker.

Introduction to Functions Test, p. 38

1. a. Two things need changed: (1) The output for Joe should not be left as a question mark but as a number.
   (2) Currently, Jane maps to two outputs: 10 and 8. One or the other row for Jane needs eliminated.

   b. No, because 5 would be mapped to both Sally and John. In a function, there needs to be exactly one output for each input, so there cannot be two outputs for 5.

2. a. Function 3. The initial value is 16.
   b. Function 2 is a linear function.
   c. Function 1: −8/3. Function 2: −1.2. Function 3: −5/3.

3. Interval −3 to −1: increasing and linear
   Interval −1 to 1.1: constant and linear
   Interval 1.1 to 1.8: decreasing and nonlinear
   Interval 1.8 to 5: increasing and nonlinear

4. a. Yes.
   b. 150 dollars per week.
   c. 200 dollars
   d. We solve the equation $150w + 200 = 1{,}200$ for w, to get $w = 100/15 = 20/3 = 6.\overline{6}$.
   After 7 weeks Leo will have saved at least $1,200.

5. a. 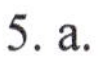

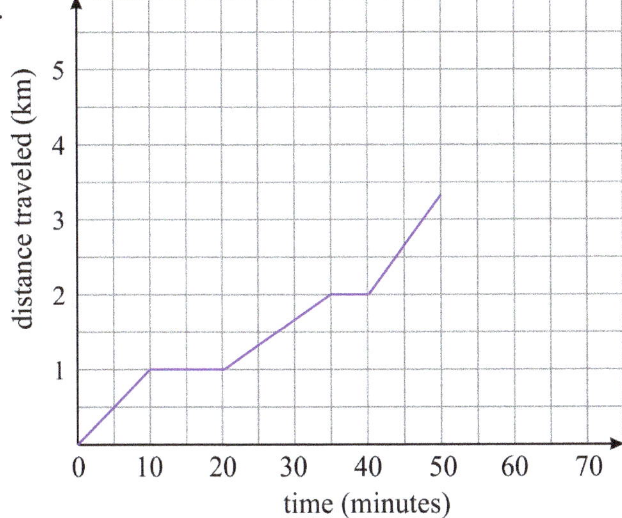

b. 3 1/3 km

6. a. 4 gal/min. It means the water is flowing to the tub at the rate of 4 gallons per minute.
b. 12 gallons. It means there were 12 gallons of water in the tub when Janet started filling it.
c. $V = 12 + 4t$ where V is the volume of water, and t is time in minutes.
d. The student graphs will vary since the scaling for the axes is not given in the problem. Check that the student's graph will show the correct line for the equation $V = 12 + 4t$. The line should go through the points (2, 20), (4, 28), and others that are on that line. For example:

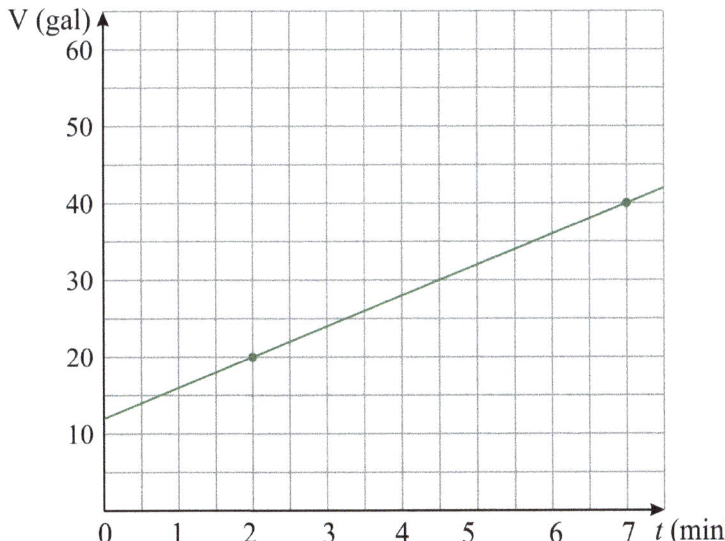

e. 52 gallons
f. In 5 minutes.

Mixed Review 5, p. 41

1. d, e, f.

2.

a. $\frac{1}{b^8}$	b. $16y^4$	c. $\frac{1}{49a^2}$	d. $-6x^9y^6$
e. $\frac{2x^6}{3}$	f. $5s^5$	g. $\frac{4x^2}{9}$	h. $\frac{a^4}{16b^8}$

3.

a. $0.0005 + 0.002 = 0.0025 = 2.5 \cdot 10^{-3}$	b. $9{,}000{,}000 + 20{,}000{,}000 = 29{,}000{,}000 = 2.9 \cdot 10^7$
c. $0.05 - 0.008 = 0.042 = 4.2 \cdot 10^{-2}$	d. $8{,}000{,}000 - 700{,}000 = 7{,}300{,}000 = 7.3 \cdot 10^6$

4. a. $1.6 \cdot 10^{17}$ b. $4 \cdot 10^2$

5. a. $h = \frac{3V}{A_b}$

b. 31.9 cm

6.

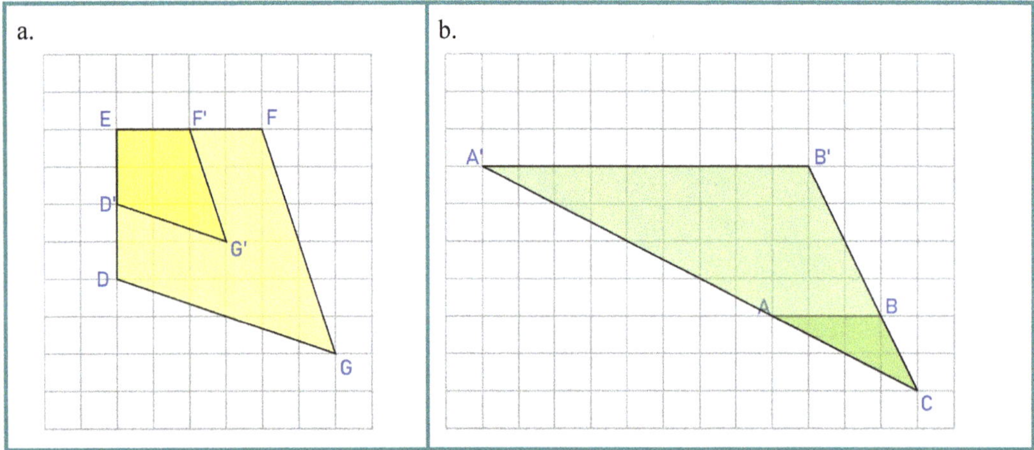

7. Let n be the number by which you multiply the current price to get the new price. Then: $n \cdot 2.76 = 2.98$, from which $n = 2.98/2.76 \approx 1.0797$, and the percent of increase is 7.97%.

Another way to solve this is to use the basic formula for the percent of increase: (difference/reference). In this case, we get: (difference/reference) = $(2.98 - 2.76)/2.76 \approx 0.0797$ or 7.97%.

8. The discounted price is $0.7 \cdot \$72.95$. Now, let t be the number by which you multiply the price to get the price with tax. Then: $t \cdot (0.7 \cdot \$72.95) = \$53.92$, from which $t = \$53.92/(0.7 \cdot \$72.95) \approx 1.056$. So, the sales tax is <u>5.6%</u>.

9.

a.

$$
\begin{aligned}
3x + \frac{x+3}{5} &= 1 && \mid \cdot 5 \\
15x + x + 3 &= 5 \\
16x + 3 &= 5 && \mid -3 \\
16x &= 2 && \mid \div 16 \\
x &= 1/8
\end{aligned}
$$

b.

$$
\begin{aligned}
3x - \frac{x+3}{5} &= 1 && \mid \cdot 5 \\
15x - (x+3) &= 5 \\
15x - x - 3 &= 5 \\
14x - 3 &= 5 && \mid +3 \\
14x &= 8 && \mid \div 14 \\
x &= 4/7
\end{aligned}
$$

Mixed Review 5, cont.

10. a. One solution.

  b. No solutions.

  c. An infinite number of solutions.

  d. One solution.

11.

a.
$$\begin{aligned} 10 + 3(a + 5) &= 2(a - 6) - 4a \\ 10 + 3a + 15 &= 2a - 12 - 4a \\ 3a + 25 &= -12 - 2a && \quad \vert\ +2a \\ 5a + 25 &= -12 && \quad \vert\ -25 \\ 5a &= -37 && \quad \vert\ \div 5 \\ a &= -37/5 \end{aligned}$$

b.
$$\begin{aligned} 20x - 2(x + 1) &= 10 - (x - 5) \\ 20x - 2x - 2 &= 10 - x + 5 \\ 18x - 2 &= 15 - x && \quad \vert\ +x \\ 19x - 2 &= 15 && \quad \vert\ +2 \\ 19x &= 17 && \quad \vert\ \div 19 \\ x &= 17/19 \end{aligned}$$

c.
$$\begin{aligned} \tfrac{1}{6}x - 1 &= 1 + \tfrac{4}{5}x && \quad \vert\ \cdot 30 \\ 30(\tfrac{1}{6}x - 1) &= 30(1 + \tfrac{4}{5}x) \\ 5x - 30 &= 30 + 24x && \quad \vert\ -5x \\ -30 &= 30 + 19x && \quad \vert\ -30 \\ -60 &= 19x && \quad \vert\ \div 19 \\ x &= -19/60 \end{aligned}$$

d.
$$\begin{aligned} 2z + \tfrac{2}{5} &= \tfrac{1}{4}z - 1 && \quad \vert\ \cdot 20 \\ 40z + 8 &= 5z - 20 && \quad \vert\ -5z \\ 35z + 8 &= -20 && \quad \vert\ -8 \\ 35z &= -28 && \quad \vert\ \div 35 \\ z &= -4/5 \end{aligned}$$

Mixed Review 6, p. 44

1.

Original figure	Dilation	Reflection
D(−5, −4)	D'(−2.5 , −2)	D"(−2.5, 2)
E(−6, −2)	E'(−3, −1)	E"(−3 , 1)
F(−1, −2)	F'(−0.5 , −1)	F"(−0.5, 1)
G(−2, −4)	G'(−1, −2)	G"(−1, 2)

2. a. See the image on the right. The vertices were originally at (−3, 4), (−3, 6), (−1, 6), and (1, 2).

Working backwards from the given, final coordinates, we get:
1. After a reflection in the x-axis: (6, −4), (6, −6), (4, −6), and (2, −2).
2. After a translation 3 units to the left: (3, −4), (3, −6), (1, −6), and (−1, −2).
3. After a rotation 180° around the origin: (−3, 4), (−3, 6), (−1, 6), and (1, 2).

b. A reflection in the vertical line $x = 1.5$.

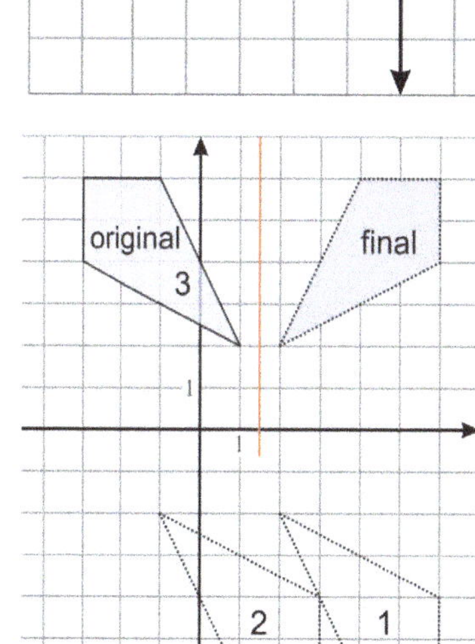

3. Let p be the price before the increases. Then we can write the equation $1.04 \cdot 1.075 \cdot 1.05p = \$136.45$, from which
$p = \$136.45/(1.04 \cdot 1.075 \cdot 1.05) = \$116.24$.
Its price had been $116.24 before the increases.

4. We cannot know whether they are similar or not.

If $\overline{AB}$ and $\overline{QR}$ were parallel, then triangles ABP and PQR would be similar, and otherwise not. But it is not stated whether $\overline{AB}$ and $\overline{QR}$ are parallel, so we cannot know whether the two triangles are similar or not.

(If $\overline{AB}$ and $\overline{QR}$ were parallel, then angle BAP and PQR would be alternate interior angles and thus congruent. Similarly, angles QRP and PBA would be alternate interior angles and congruent. Angles BPA and QPR are congruent, being vertical angles. The triangles ABP and PRQ would have three congruent angles, and thus would be similar.)

5. Angle ABC is supplementary to the 137° angle, so it measures 43°. In triangle ABC, $y = 180° - 82° - 43° = 55°$. Angle x is an alternate interior angle with angle y, so it is congruent with y and also measures 55°.

6. The volume of the bucket, in cubic inches, is $V = \pi\,(5.75 \text{ in})^2 \cdot 11 \text{ in} \approx 1142.558$ cubic inches. In gallons, this is $1142.558 \text{ in}^3 \cdot 1 \text{ gal}/(231 \text{ in}^3) \approx 4.9$ gallons. Yes, four gallons of water fit in the bucket and you should not believe your friend.

7. a. 230 cm = <u>91</u> in b. 54 L = <u>14</u> gal

c. 24.5 ft = <u>7.47</u> m d. 437 in = <u>11.1</u> m

8. $\dfrac{1 \cdot 10^2 \text{ g}}{2.99 \cdot 10^{-23} \text{ g}} \approx 0.334448 \cdot 10^{25}$ atoms $\approx 3.34 \cdot 10^{24}$ atoms.

9. a. After multiplying both sides by 2, the term x gets forgotten.
b. When the term $-(x - 1)/5$ is multiplied by 5, since there is a minus sign in front of it, it should become $-x + 1$.

a.
$$\begin{aligned} 5 + x + \frac{x-3}{2} &= 6 & &\Big|\ \cdot 2 \\ 10 + 2x + x - 3 &= 12 \\ 7 + 3x &= 12 & &\Big|\ -7 \\ 3x &= 5 & &\Big|\ \div 3 \\ x &= 5/3 = 1\ 2/3 \end{aligned}$$

b.
$$\begin{aligned} x - 2 &= 2x - \frac{x-1}{5} & &\Big|\ \cdot 5 \\ 5x - 10 &= 10x - x + 1 \\ 5x - 10 &= 9x + 1 & &\Big|\ -9x \\ -4x - 10 &= 1 & &\Big|\ +10 \\ -4x &= 11 & &\Big|\ \div(-4) \\ x &= -11/4 = -2\ 3/4 \end{aligned}$$

10. a. $a_2 = 4m - a_1 - a_3 - a_4$

b. At least 92 points.

11. Let x be the first such even whole number. Then the other two are $x + 2$ and $x + 4$. We get the equation $x + x + 2 + x + 4 = 13{,}788$, from which $3x = 13{,}782$ and $x = 4{,}594$. The three numbers are 4,594, 4,596, and 4,598.

Graphing Linear Equations Review, p. 47

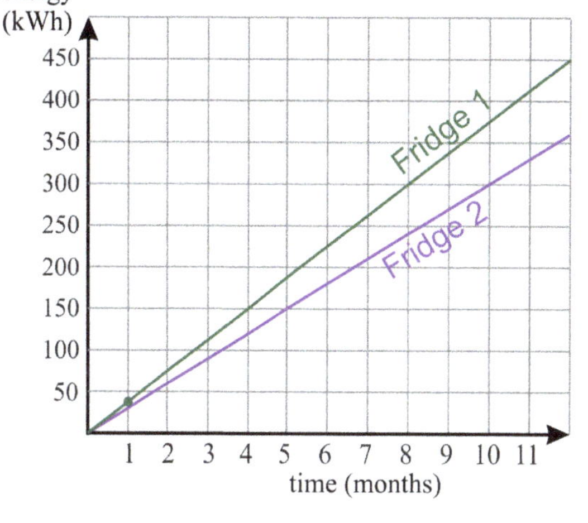

1. a. Fridge 1. From the graph, we can see that Fridge 2 consumes 150 kWh in 5 months, which means it consumes 30 kWh/month. Fridge 1 consumes 37.5 kWh per month. So, Fridge 1 consumes <u>7.5 kWh more</u> in a month than Fridge 2.

   b. Fridge 1: $E = 37.5t$. Fridge 2: $E = 30t$.

   c. and d. See the image on the right.

2. a. The brown lines: $y = -2x$ and $y = -2x + 10$

   The green lines: $y = (1/2)x$ and $y = (1/2)x + 5/2$

   b. The easiest way is to draw an outer rectangle around it, calculate the areas of the triangles marked with 1, 2, 3, and 4, and lastly subtract the areas of those triangles from the area of the outer rectangle.

   The area of the outer rectangle is 5 by 4 units = 20 square units.
   The areas of triangles 1, 2, 3, and 4 are 4, 1, 4, and 1 square units, respectively.
   The area of the inner rectangle is therefore 20 − 4 − 1 − 4 − 1 = <u>10 square units</u>.

3. a. $y = (3/4)x + 9/2$ b. $y = -10$

4. a. This line passes through the points (5, 12) and (25, 3). The slope is therefore −9/20 or −0.45.
   b. This line passes through the points (10, 6) and (25, 9). The slope is therefore 3/15 = 1/5.

   Equations:

   a. To find the y-intercept, we substitute (5, 12) into the equation $y = -9/20x + b$, and solve the resulting equation (see the solution on the right).

   $$\begin{aligned} 12 &= -9/20(5) + b \\ 12 &= -9/4 + b \quad \Big| + 9/4 \\ b &= 12 + 9/4 \\ b &= 14\tfrac{1}{4} \text{ or } 14.25 \end{aligned}$$

   The final equation is $y = -0.45x + 14.25$

   b. To find the y-intercept, we substitute (10, 6) into the equation $y = (1/5)x + b$, and solve the resulting equation (see the solution on the right).

   $$\begin{aligned} 6 &= (1/5)10 + b \\ 6 &= 2 + b \\ b &= 4 \end{aligned}$$

   The equation is $y = (1/5)x + 4$.

5. We calculate the slope using any two of the three points, and check whether we get the same result.

   Using the first two points, the slope is −5/2. Using the last two points, the slope is −4/2 = −2.
   Using the first and the last point, the slope is −9/4. So, these three points do not fall on the same line.
   Encourage the student to also plot the points, and to see the slight difference in the slopes between the points.

6. Using the two known points, we can calculate the slope. It is 9/12 = 3/4. The equation of the line is of the form $y = (3/4)x + b$. We substitute (3, 9) in it: $9 = (3/4)(3) + b$. From that, $b = 27/4$. So, the equation is $y = (3/4)x + 27/4$.

   Now, the point (s, 12) also fulfills that equation, so we can write $12 = (3/4)s + 27/4$, and from that solve that $\underline{s = 7}$.

7. a. The equation of line S is $y = (6/5)x + 4$. The slope of line T is −5/6. Since it passes through (1, 1), we can substitute those coordinates in the equation $y = (-5/6)x + b$ to get $1 = -5/6 + b$, from which $b = 11/6$.

   So, the equation of line T is $y = (-5/6)x + 11/6$.

   b. To transform this to the standard form, we multiply it by 6: $6y = -5x + 11$, from which we get $\underline{5x + 6y = 11}$.

8. It is a linear relationship. Equation: $C = 110d + 1{,}500$, where C is the cost of running the truck and d is the number of days.

9.

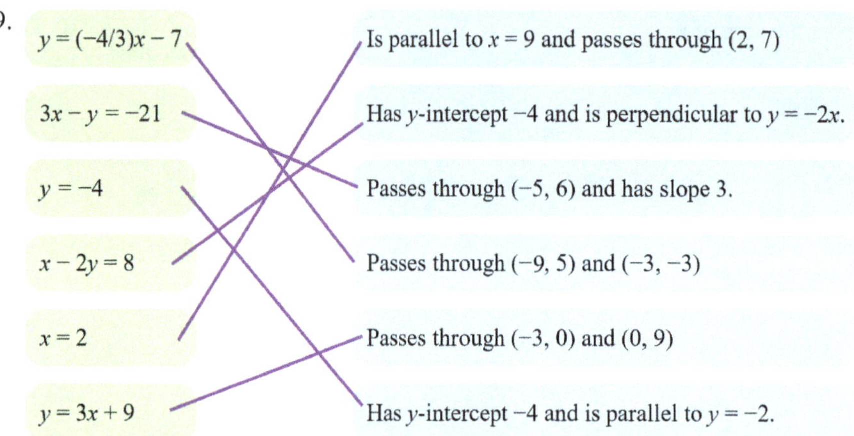

10. a. $2x - y = -10$. x-intercept: −5, y-intercept: 10.
    b. $2x + 9y = -6$. x-intercept: −3, y-intercept: −2/3.

11. a. $T = (20/3)t - 5$, where T is temperature, and t is time in hours.
    b. 2:30 PM is 4.5 hours after 10 AM. Using our equation, $T = (20/3)(4.5) - 5 = 25°C$.

    c.
$$\begin{aligned} 22 &= (20/3)t - 5 && \cdot 3 \\ 66 &= 20t - 15 && +15 \\ 81 &= 20t && \div 20 \\ 81/20 &= t \\ t &= 4.05 \text{ or } 4\ 1/20 \end{aligned}$$

    4 1/20 hours is 4 hours 3 minutes. The temperature reaches 22°C at 2:03 PM.

    d. 1:45 is 3.75 hours after 10 AM. $T = (20/3)(3.75) - 5 = \underline{20°C}$

    e. The equation for this changed situation is simply $T = (20/3)t - 12$. We solve the equation:

$$\begin{aligned} 22 &= (20/3)t - 12 && \cdot 3 \\ 66 &= 20t - 36 && +36 \\ 102 &= 20t && \div 20 \\ 102/20 &= t \\ t &= 5.1 \text{ or } 5\ 1/10 \end{aligned}$$

    5.1 hours is 5 hours 6 minutes. The house reaches the temperature of 22°C at 3:06 PM.

Graphing Linear Equations Test, p. 51

1. a. The slope is 400 m^2/(25 L) = 16 m^2/L.

   b. Paint 2. Paint 2 covers 16 m^2 per liter, whereas Paint 1 only covers 12 m^2 per liter.

   c. Paint 1: 30 m^2/(12 m^2/L) = 2.5 liters.
   Paint 2: 30 m^2/(16 m^2/L) = 1.875 liters.

   d. See the graph on the right.

   e. For one liter, Paint 2 covers 4 m^2 more than Paint 1. So, for 40 liters, Paint 2 covers 40 L · 4 m^2 = <u>160 m^2</u> more than Paint 1.

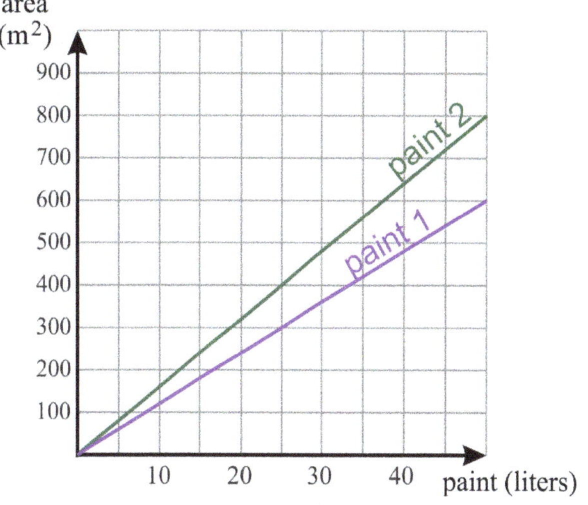

2. a. $y = (3/4)x - 8$
   b. $y = (-5/2)x - 2$
   c. $x = -11$
   d. $y = 6x - 22$

3. a. W = 7.6 − 0.4t, where W is the weight in kilograms, and t is the time in months.
   b. The answers will vary since the scaling on the vertical axis will vary. Check the student's graph. For example:

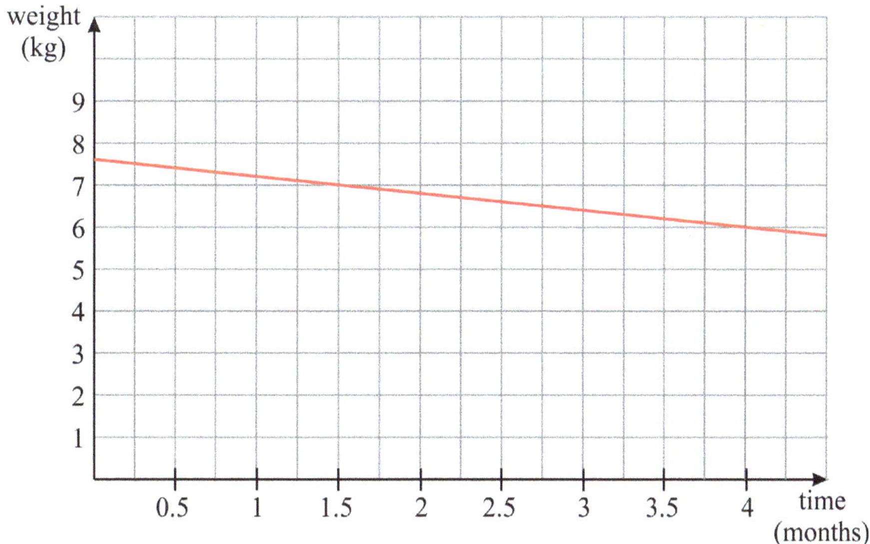

   Here is another possibility, where the vertical axis does not start from zero:

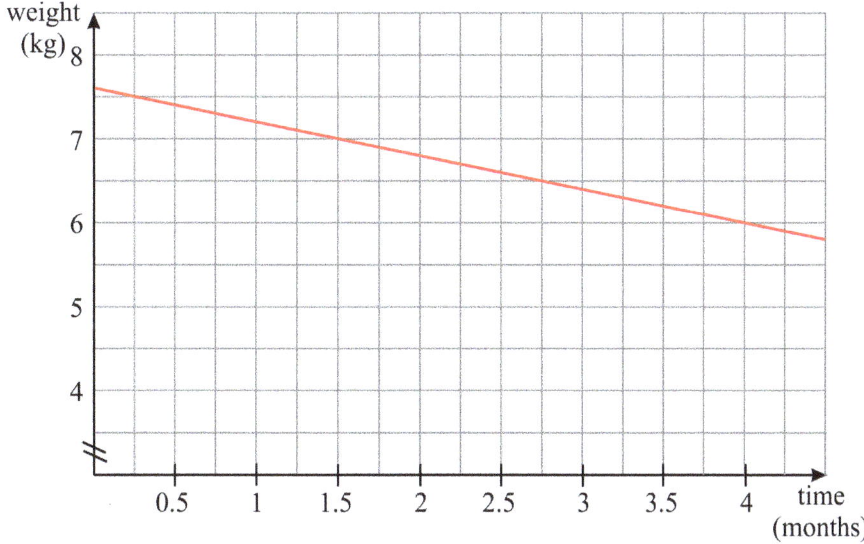

   c. Solving the equation 5 = 7.6 − 0.4t, we get t = 2.6/0.4 = <u>6.5 months</u>.

Graphing Linear Equations Test, cont.

4. a. This line goes through the points (0, 10) and (15, 30), from which we can calculate the slope as 20/15 = 4/3. It crosses the y-axis at $y = 10$, so the equation is $y = (4/3)x + 10$.

   b. Here we cannot easily see the y-intercept. But we can find two points the line goes through: (−5, 0) and (2.5, −20). From those, we can calculate the slope as −20/7.5 = −20/(15/2) = −40/15 = −8/3.

   So, the equation is of the form $y = (-8/3)x + b$. Now we will substitute one of the points on the line, such as (−5, 0), into that equation, to find the value of b. We get $0 = (-8/3)(-5) + b$, from which $b = -40/3$. So, the equation is $\underline{y = (-8/3)x - 40/3}$. If the student gave it in standard form, it is $8x + 3y = -40$.

5. The equation is $y = 3x - 2$. See the graph on the right.

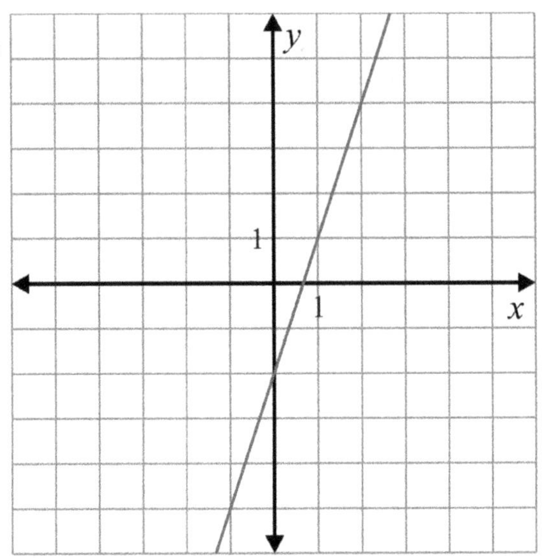

6. a. The slope of Line L is 7/(−5) = −7/5. Its equation is $y = (-7/5)x + b$. Substituting (2, −1) into that equation, we get $-1 = (-7/5)(2) + b$, from which $b = 9/5$. So, the equation is $y = (-7/5)x + 9/5$.

   Since M is perpendicular to L, its slope is 5/7. The equation of M is of the form $y = (5/7)x + b$. We substitute (−2, −3) into that, and get: $-3 = (5/7)(-2) + b$. From that, $b = -3 + 10/7 = -11/7$. So, the equation of line M is $y = (5/7)x - 11/7$.

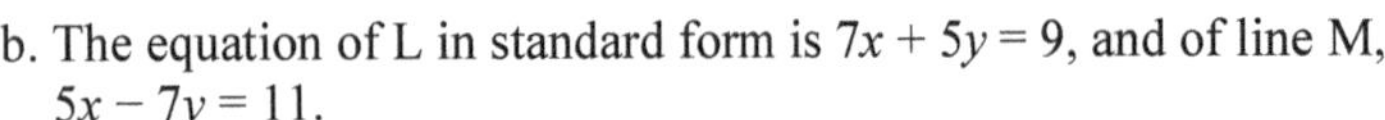

   b. The equation of L in standard form is $7x + 5y = 9$, and of line M, $5x - 7y = 11$.

7. The slope of this line is (−28 − (−1))/(10 − 1) = −27/9 = −3. For (6, s) to be on the same line, the slope calculated using the points (1, −1) and (6, s) needs to be −3. Writing that as an equation, we get $(s - (-1))/(6 - 1) = -3$. See its solution below:

$$\begin{aligned} \frac{s-(-1)}{6-1} &= -3 \\ \frac{s+1}{5} &= -3 \qquad \Big| \cdot 5 \\ s+1 &= -15 \\ s &= -16 \end{aligned}$$

   So, the solution is that $s = -16$.

   You can also solve this in this manner. After finding out that the slope is −3, we can advance from the point (1, −1), adding one to the x-value and subtracting 3 from the y-value each time, until we come to the x-value of 6: (1, −1) → (2, −4) → (3, −7) → (4, −10) → (5, −13) → (6, −16).

Mixed Review 7, p. 55

1.

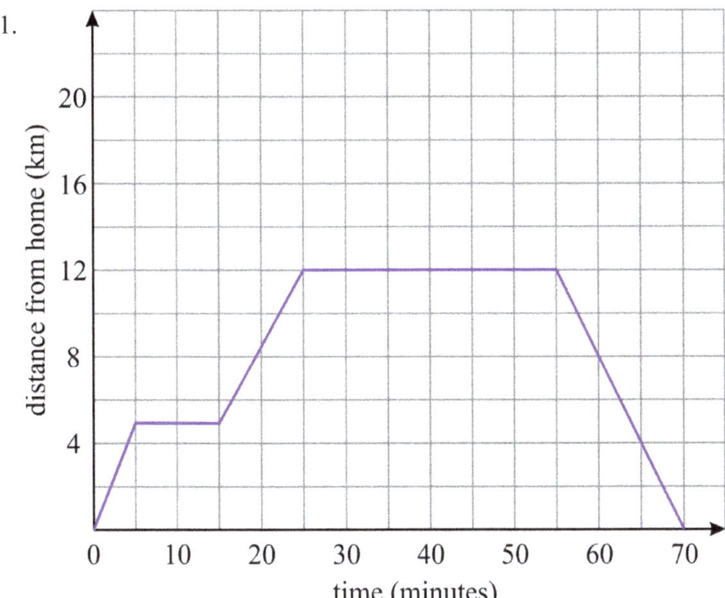

2. Answers will vary because there are many possible sequences of transformations that will map figure 3 onto figure 4. Check the student's answer. For example: Reflect Figure 3 in the vertical line at $x = 3$, and then translate it six units up and one unit to the right. Or: First, translate it six units up, then reflect it in the vertical line at $x = 3.5$.

3. a. It has no solutions.
   b. Answers will vary; check the student's answer. For example, $3(y - 1) = 7 + 2y$ or $4(y - 1) = 7 + 3y$.
   After the change, if there is a term with y on both sides, those terms should not have the same coefficient.

4. a. Yes, they are. The two triangles have the same angles, so they are similar triangles. First, both triangles have one right angle. Secondly, the angles BCA and DCE are congruent because they are vertical angles.

   Lastly, the third angles of both triangles must be congruent. To see that, recall that the sum of the angles in any triangle is 180°. So, ∠BAC = 180° − 90° − ∠BCA and ∠EDC = 180° − 90° − ∠DCE. But ∠BCA and ∠DCE are congruent, therefore ∠BAC and ∠EDC end up being congruent also.

   b. The side BC corresponds to the side CE = 9.6 ft in triangle CED. We can write a proportion:
   $x/5.3 = 9.6/10.3$, from which $x = 5.3 \cdot (9.6/10.3) \approx 4.939806$. So, <u>$x$ is 4.9 ft</u>.
   (The value of x can be found in other ways, too.)

5. a. <u>Function 3</u>. Its rate of change in the interval [1, 3] is (6 − (−4)/2 = 5. The rate of change of Function 2 is 2. The rate of change of Function 1 in this interval is negative.

   b. <u>Function 3</u>. In the interval [4, 5], the rate of change of Function 1 is (2.5 − (−1.7)/1 = 4.2. The rate of change of Function 2 is 2, and of Function 3 is 5.

Mixed Review 7, cont.

6.

a.
$$\begin{aligned} 10(x-3)+2x-5 &= 6-3x \\ 10x-30+2x-5 &= 6-3x \\ 12x-35 &= 6-3x \\ 15x &= 41 \\ x &= 41/15 \end{aligned}$$

b.
$$\begin{aligned} \frac{1}{3}x-5 &= \frac{1}{4}x+2 \quad \Big| \cdot 12 \\ 4x-60 &= 3x+24 \\ x-60 &= 24 \\ x &= 84 \end{aligned}$$

c.
$$\begin{aligned} \frac{1}{6}(x-7) &= -\frac{7}{8} \quad \Big| \cdot 24 \\ 4(x-7) &= -21 \\ 4x-28 &= -21 \\ 4x &= 7 \\ x &= 7/4 \end{aligned}$$

d.
$$\begin{aligned} \frac{5x-2}{10}-2 &= 3x \quad \Big| \cdot 10 \\ (5x-2)-20 &= 30x \\ 5x-22 &= 30x \\ -22 &= 25x \\ x &= -22/25 \end{aligned}$$

7. Let M be Mary's age now, and R be Ryan's age now. Then, the equations we get are:

$$\begin{cases} M-20 = (3/5)(R-20) & \Big| \cdot 5 \\ M = (7/9)R & \Big| \cdot 9 \end{cases}$$

Substituting $M = (7/9)R$ into the first equation, we get:

$$\begin{aligned} (7/9)R-20 &= (3/5)(R-20) \quad \Big| \cdot 45 \\ 35R-900 &= 27(R-20) \\ 35R-900 &= 27R-540 \\ 8R &= 360 \\ R &= 45 \end{aligned}$$

Then, $M = (7/9)R = (7/9)45 = 35$.

Solution: Mary is 35 years old and Ryan is 45 years old.

8. Let n be the number of nickels in her piggy bank. Then, the number of dimes is $2n$. Since the value of her coins is \$12.25, we can write the equation:

$$\begin{aligned} 5n+10\cdot 2n+16\cdot 25 &= 1225 \\ 5n+20n+400 &= 1225 \\ 25n+400 &= 1225 \\ 25n &= 825 \\ n &= 33 \end{aligned}$$

So, she has 33 nickels, 66 dimes, and 16 quarters.

Mixed Review 8, p. 58

1. It will have no solutions if the coefficient of the x-term is the same on both sides of the equation, and if the constant terms will not cancel out. We can simplify both sides of the equation (separately) to get:

$4x + 28 = 10 - 2ax + 10$
$4x + 28 = -2ax + 20$

The constant terms on the two sides are different (28 and 20). So, when $-2a = 4$, or $a = -2$, the equation will have no solutions.

2\. a.

$$A = \frac{pq}{2} \quad \Big| \cdot 2$$

$$2A = pq \quad \Big| \div q$$

$$\frac{2A}{q} = p$$

Solution: $p = (2A)/q$

b. $p = (2 \cdot 0.6\text{ m}^2)/(0.9\text{ m}) = (1.2\text{ m}^2)/(0.9\text{ m}) = 12/9\text{ m} = 4/3\text{ m}$, or about <u>133 cm</u>.

3\. a. $V = \pi(4.5\text{ in})^2 \cdot 4\text{ in} \div 12 \approx 21\text{ in}^3$

b. We first need to calculate the full volume of the 9-inch diameter circular cake: $\pi(4.5\text{ in})^2 \cdot 4\text{ in} \approx 254\text{ in}^3$. Let h be the height of the square cake. We can write the equation $8\text{ in} \cdot 8\text{ in} \cdot h = 254\text{ in}^3$, from which $h = 254/64$ inches, or about 4 inches. So both cakes need to be about the same height to have the same volume.

4\.

a.	b.
$2x = \frac{x-5}{6} - 2 \quad \Big\| \cdot 6$	$\frac{x-1}{4} + \frac{2x+7}{3} = 0 \quad \Big\| \cdot 12$
$12x = x - 5 - 12$	$3(x-1) + 4(2x+7) = 0$
$11x = -17 \quad \Big\| \div 11$	$3x - 3 + 8x + 28 = 0$
$x = -17/11$	$11x = -25 \quad \Big\| \div 11$
	$x = -25/11$

5\. a. $d = 600t$

b. 600 mph

c. Yes, because it is of the form $y = mx$ (or, dependent variable = constant times the independent variable.) Here, m is 600.

d. Student graphs will vary because the scaling on the d-axis may vary. For example:

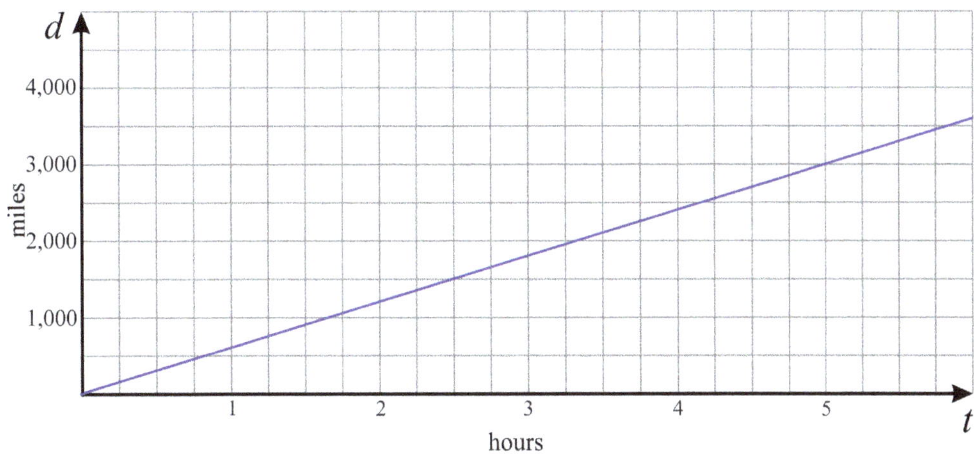

e. The time 1 hr 40 minutes needs to be in hours before we can use the equation. 1 hr 40 min = 1 2/3 hr = 5/3 hr. Then, $d = 600(5/3) =$ <u>1,000 miles</u>.

6\. a. No. While the time increases steadily by half an hour each time, the distance does not increase by the same amount each time. For example, from 0.5 to 1 hour, the distance increases by 300 miles, but from 1 to 1.5 hours, it increases by 350 miles.

b. The function in question #5 has a constant rate of change of 600 mph at all times.
The function in question #6 has a rate of change of (2,400 mi − 1,750 mi)/1 hr = 650 mph.
<u>The function in #6 has a larger rate of change from 3 to 4 hours</u>.

Mixed Review 8, cont.

7.

a. $(2x^3)^{-1} = \dfrac{1}{2x^3}$	c. $(ab^3)^{-2} = \dfrac{1}{a^2b^6}$	e. $(-2x)^5 \cdot (5x)^{-1} = \dfrac{-32x^4}{5}$
b. $(8a^{-5})^2 = \dfrac{64}{a^{10}}$	d. $(-3s^{-2}t)^3 = \dfrac{-27t^3}{s^6}$	f. $z^4 \cdot (-3z^3)^{-2} = \dfrac{1}{9z^2}$

8. a. From $x = -25$ to $x = -20$
   b. From $x = -10$ to $x = -2.5$
   c. From $x = -2.5$ to $x = 7.5$

9. Answers will vary. Check the student's graph. For example:

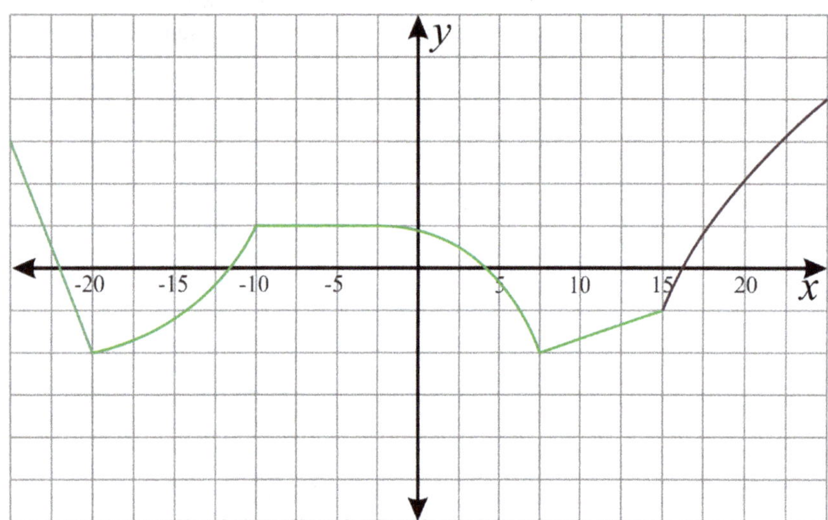

Irrational Numbers and the Pythagorean Theorem Review, p. 61

1. a. 8 b. 13 c. 50 d. 0.9
   e. 6/10 = 3/5 f. −7 g. 5 h. 30

2. a. between 2 and 3 b. between 8 and 9 c. between 11 and 12 d. between 6 and 7

3. Here is one possible way the process could go. The intermediate guesses by the student(s) will vary. The table is just showing one example.

$\sqrt{13} \approx 3.6$

Low Guess	$(LG)^2$	$(HG)^2$	High Guess
3.6	12.96	13.3225	3.65
3.6	12.96	13.0321	3.61

4.

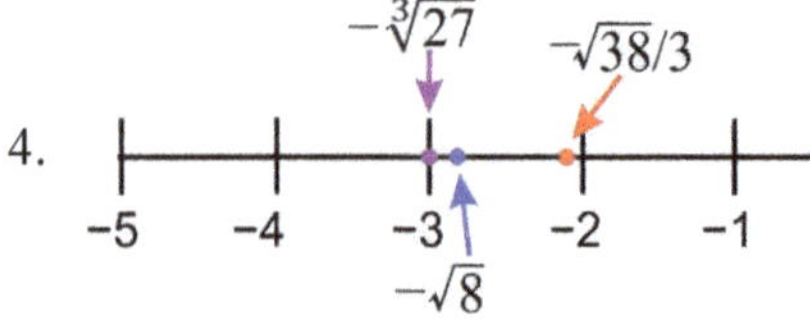

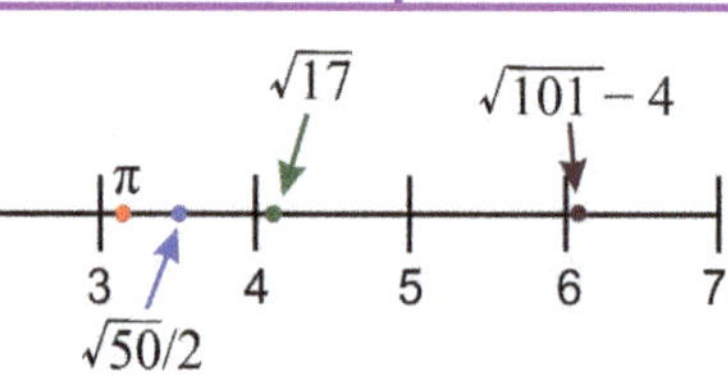

5.

a. $11 < \sqrt{150}$	b. $\sqrt{76} < 9$	c. $\sqrt{20} > 4$	d. $\sqrt[3]{10} > 2$
e. $4 < \pi + 1$	f. $\sqrt{85}/3 > 3$	g. $\sqrt{27} + 2 > 6$	h. $\sqrt{68} - 3 < 6$

6. a. 12 b. −9 c. 40
   d. 8 e. 49 f. 20

7. a. 7 square units
   b. $4\sqrt{20}$ (or this can also be written as) $8\sqrt{5}$)

8. a. rational; it is a decimal that terminates
   b. rational; the value is 50
   c. irrational; π is irrational, 5 is rational, and an irrational number times a rational is irrational
   d. irrational; 56 is not a perfect square, so $\sqrt{56}$ is irrational and so is its opposite
   e. rational; this is a ratio of two integers
   f. rational; the value is 1/6
   g. rational; this is a repeating decimal
   h. rational; this is a terminating decimal
   i. rational; this is a repeating decimal
   j. irrational; Seven is not a perfect square so $\sqrt{7}$ is irrational. Four is rational, and an irrational number times a rational number ($4\sqrt{7}$) is irrational.

9. a.

$$\begin{aligned} 100x &= 61.6161\ldots \\ -\ x &= 0.6161\ldots \\ \hline 99x &= 61 \\ x &= \underline{61/99} \end{aligned}$$

b.

$$\begin{aligned} 10x &= 41.7777\ldots \\ -\ x &= 4.1777\ldots \\ \hline 9x &= 37.6 \\ x &= 37.6/9 = 376/90 = \underline{188/45} \end{aligned}$$

10. a. $x = \sqrt{147}$ or $x = -\sqrt{147}$
    b. $a = 13$ or $a = -13$
    c. $w = \sqrt[3]{0.36}$
    d. $x = \sqrt[3]{7}$
    e. $b = 5$
    f. $a = -2$

11. a.

$$\begin{aligned} y^2 + 18 &= 35 \\ y^2 &= 17 \\ y &= \sqrt{17} \approx 4.123 \\ \text{or } y &= -\sqrt{17} \approx -4.123 \end{aligned}$$

Check: $(\sqrt{17})^2 + 18 \stackrel{?}{=} 35$

$17 + 18 = 35$ ✓

b.

$$\begin{aligned} 0.6h^2 &= 4 \\ h^2 &= 4/0.6 = 40/6 = 20/3 \\ h &= \sqrt{20/3} \approx 2.582 \\ \text{or } h &= -\sqrt{20/3} \approx -2.582 \end{aligned}$$

Check: $0.6 \cdot (\sqrt{20/3})^2 \stackrel{?}{=} 4$

$0.6 \cdot (20/3) \stackrel{?}{=} 4$

$(6/10) \cdot (20/3) \stackrel{?}{=} 4$

$120/30 = 4$

12. a.

$$\begin{aligned} 20^2 + 24^2 &\stackrel{?}{=} 30^2 \\ 400 + 576 &\stackrel{?}{=} 900 \\ 976 &> 900 \end{aligned}$$

No, they don't form a right triangle. (They would form an acute triangle.)

b.

$$\begin{aligned} 1^2 + 2.4^2 &\stackrel{?}{=} 2.6^2 \\ 1 + 5.76 &\stackrel{?}{=} 6.76 \\ 6.76 &= 6.76 \end{aligned}$$

Yes, they form a right triangle.

13. We can ignore the negative answers because a side cannot have a negative length.

a.

$$\begin{aligned} s^2 &= 3^2 + 5^2 \\ s^2 &= 9 + 25 \\ s^2 &= 34 \\ s &= \sqrt{34} \end{aligned}$$

b.

$$\begin{aligned} y^2 + 12^2 &= 14^2 \\ y^2 + 144 &= 196 \\ y^2 &= 52 \\ y &= \sqrt{52} \end{aligned}$$

14.

$$\begin{aligned} x^2 + 21.1^2 &= 22.5^2 \\ x^2 + 445.21 &= 506.25 \\ x^2 &= 61.04 \\ x &= \sqrt{61.04} \approx \underline{7.8\text{ m}} \end{aligned}$$

15. Let x be the hypotenuse of this triangle.

$$\begin{aligned} (\sqrt{7})^2 + (\sqrt{8})^2 &= x^2 \\ 7 + 8 &= x^2 \\ x^2 &= 15 \\ x &= \sqrt{15} \end{aligned}$$

The hypotenuse is $\sqrt{15}$ units long.

16. The pennant is an isosceles triangle.

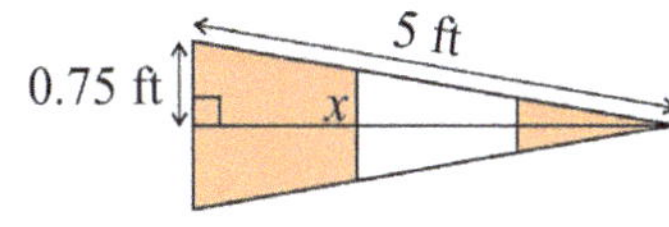

We calculate its altitude using the Pythagorean Theorem. From the right triangle in the image, we get:

$$\begin{aligned} 0.75^2 + x^2 &= 5^2 \\ 0.5625 + x^2 &= 25 \\ x^2 &= 24.4375 \\ x &= \sqrt{24.4375} \approx 4.94343\ldots\text{ ft} \end{aligned}$$

So the area is $A = bh/2 \approx 1.5\text{ ft} \cdot 4.94343\text{ ft}/2 = 3.70757\text{ ft}^2 \approx \underline{3.7\text{ ft}^2}$.

17. The figure below has four right triangles, each with sides *a, b* and *c*. The sides of the outside square are $a + b$. The triangles enclose a square with sides c units long.

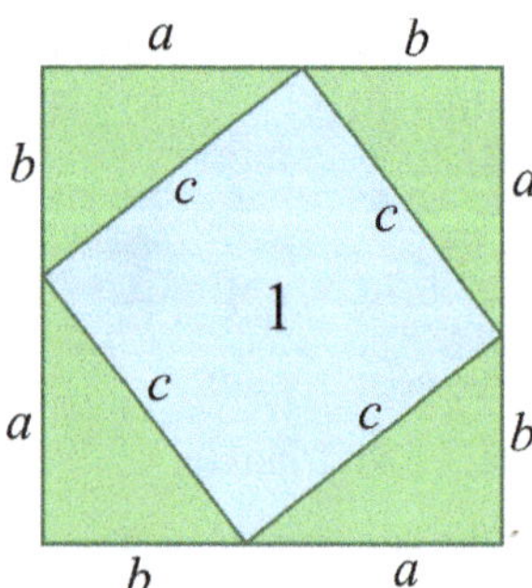

Below, the sides of the large square are still $a + b$, but but the four right triangles have been rearranged so that two smaller squares are formed, with sides a and b.

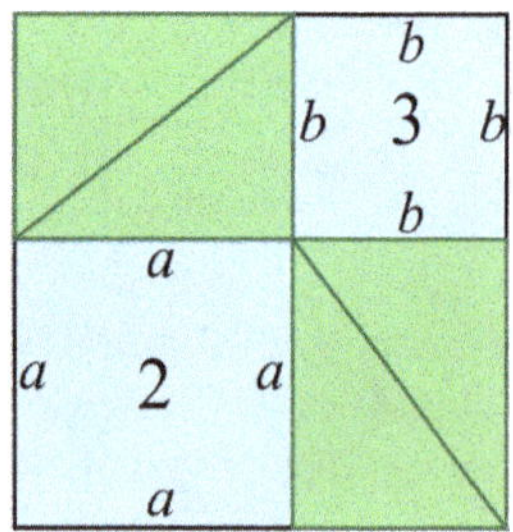

Since the areas of both large squares are equal, and the areas of the four right triangles are equal, it follows that the remaining (blue) areas are also equal. In other words, the area of square 1, which is c^2, equals the area of square 2 (which is a^2) plus the area of square 3 (which is b^2). In symbols, $c^2 = a^2 + b^2$. ☺

18. This is a trapezoid. Its area is the average length of the two parallel sides, times its altitude. The altitude is 5 units. The two parallel sides measure 5 and 9 units. So, the area is $(5 + 9)/2 \cdot 5 = 7 \cdot 5 = 35$ square units.

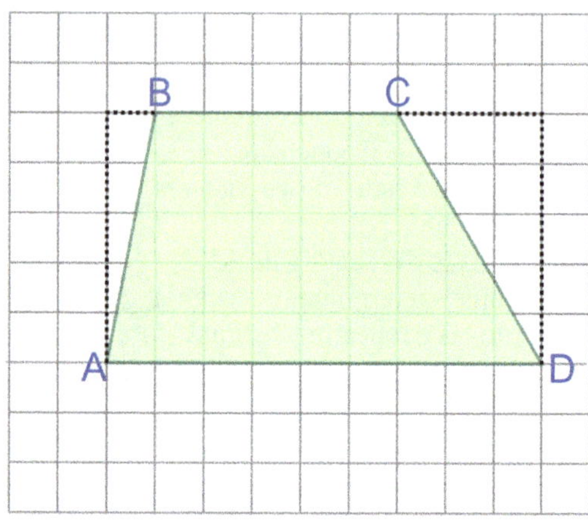

For the perimeter, we will find the side lengths AB and CD using the Pythagorean Theorem. The image shows with dashed lines the right triangles we will use.

For AB:
$$\begin{aligned} 5^2 + 1^2 &= AB^2 \\ 26 &= AB^2 \\ AB &= \sqrt{26} \approx 5.099 \text{ units} \end{aligned}$$

For CD:
$$\begin{aligned} 3^2 + 5^2 &= CD^2 \\ 34 &= CD^2 \\ CD &= \sqrt{34} \approx 5.831 \text{ units} \end{aligned}$$

The perimeter is then the sum of all four sides:
$P = 5.099 + 5 + 5.831 + 9 = 24.93$ units.
Since each unit is 2.0 feet, the perimeter is 49.86 feet ≈ <u>50 feet</u>.

19. a. The horizontal distance between the points is $|11 - (-3)| = 14$ and the vertical distance is $|-9 - 4| = 13$.

$$\begin{aligned} x^2 &= 14^2 + 13^2 \\ x^2 &= 196 + 169 \\ x &= \sqrt{365} \approx 19.1 \text{ units} \end{aligned}$$

b. The horizontal distance between the points is $|70 - 42| = 28$ and the vertical distance is $|100 - (-15)| = 115$.

$$\begin{aligned} x^2 &= 28^2 + 115^2 \\ x^2 &= 14{,}009 \\ x &= \sqrt{14{,}009} \approx 118.4 \text{ units} \end{aligned}$$

Irrational Numbers and the Pythagorean Theorem Review, cont.

20. Please see the image below:

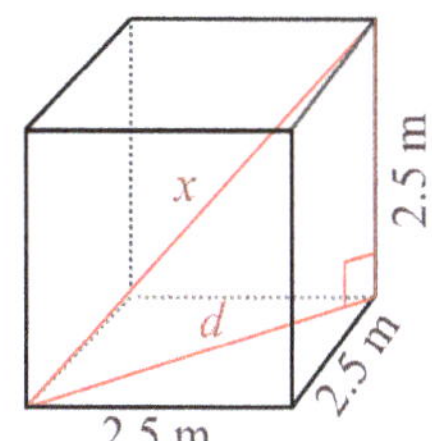

We will first solve for d using the right triangle on the bottom of the cube:

$$2.5^2 + 2.5^2 = d^2$$
$$12.5 = d^2$$
$$d = \sqrt{12.5}$$

Next, we will use the Pythagorean Theorem again in the pink triangle:

$$(\sqrt{12.5})^2 + 2.5^2 = x^2$$
$$12.5 + 6.25 = x^2$$
$$x = \sqrt{18.75}$$
$$x \approx \underline{4.3\text{ m}}$$

21. Let x be the distance from B to C along 5th Avenue South.

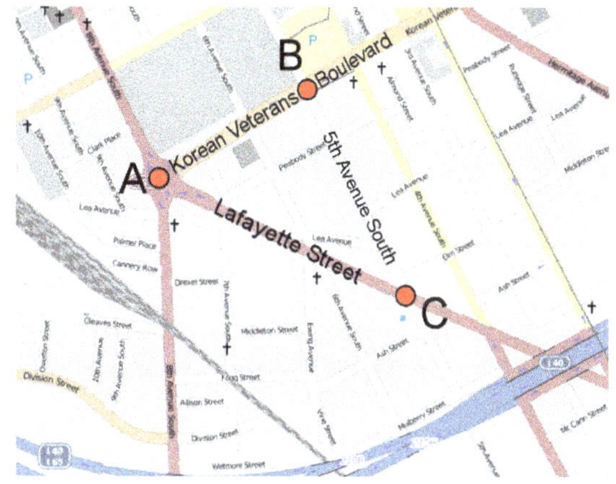

Then:

$$370^2 + x^2 = 620^2$$
$$136{,}900 + x^2 = 384{,}400$$
$$x^2 = 247{,}500$$
$$x = \sqrt{247{,}500} \approx 497.49\text{ m} \approx 500\text{ m}$$

To go directly from A to C is 620 m, and the distance from A to B and then to C is 370 m + 500 m = 870 m. Therefore, to go directly from A to C is 870 m − 620 m = <u>250 m shorter</u> than to go from A to B and then to C.

Irrational Numbers and the Pythagorean Theorem Test, p. 67

1. (b), (c), and (d) are correct. The statement (a) is almost correct. It is true that 0.141414 is rational, but not because its decimal expansion repeats. Its decimal expansion does not actually repeat indefinitely anyway; its decimal expansion terminates. The reason it is rational is because we can write it as a ratio of two whole numbers: 141,414/1,000,000.

2. a. $6 < \sqrt{44} < 7$ b. $-3 < -\sqrt{5} < -2$ c. $2 < \sqrt[3]{21} < 3$

3.

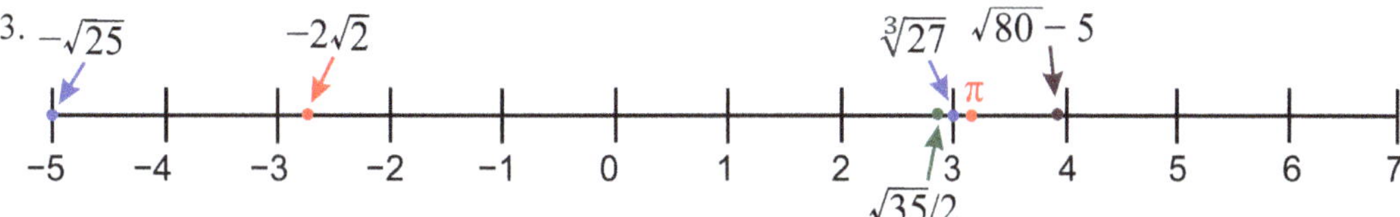

4.

a. $x^2 = 37$ $x = \sqrt{37}$ or $x = -\sqrt{37}$	b. $5a^2 = 90$ $a^2 = 18$ $a = \sqrt{18}$ or $a = -\sqrt{18}$	c. $w^3 = 125$ $w = 5$

5.

$$\begin{aligned} 100x &= 23.232323\ldots \\ -\ x &= 0.232323\ldots \\ \hline 99x &= 23 \\ x &= \underline{23/99} \end{aligned}$$

6. Let x be the hypotenuse. Using the Pythagorean Theorem, we get:

$$\begin{aligned}(\sqrt{21})^2 + (\sqrt{15})^2 &= x^2\\ 21 + 15 &= x^2\\ x^2 &= 36\\ x &= 6\\ \text{or } x &= -6\end{aligned}$$

Since this is a triangle, we omit the negative solution. The hypotenuse is 6 units long.

7. The horizontal distance between the two points comes from their x-coordinates and is $|10 - (-2)| = 12$ units. The vertical distance comes from the y-coordinates and is $|8 - 5| = 3$ units long. Let x be the desired distance between the two points. Using the Pythagorean Theorem:

$$\begin{aligned}12^2 + 3^2 &= x^2\\ 144 + 9 &= x^2\\ 153 &= x^2\\ x &= \sqrt{153} \approx 12.37 \text{ units}\end{aligned}$$

8. In the image, there is a right triangle with 0.3 km and 0.5 km legs. Its hypotenuse is marked with x. Applying the Pythagorean Theorem to that triangle, we get:

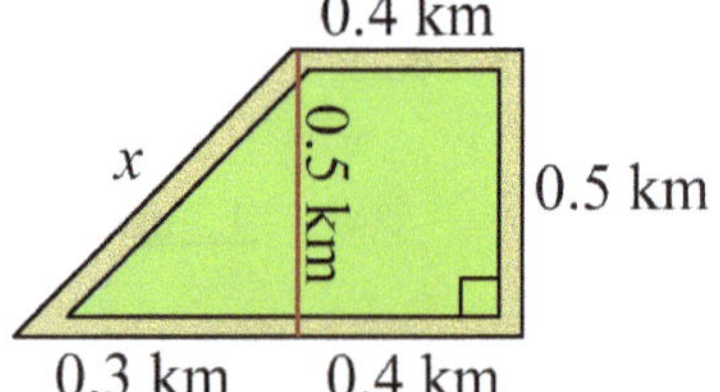

$$\begin{aligned}x^2 &= 0.3^2 + 0.5^2\\ x^2 &= 0.09 + 0.25\\ x^2 &= 0.34\\ x &= \sqrt{0.34} \approx 0.6 \text{ km}\end{aligned}$$

Now we can calculate the perimeter of the track: it is 0.4 km + 0.5 km + 0.7 km + 0.6 km = 2.2 km. When she runs around it three times, she runs a total distance of 6.6 km.

Mixed Review 9, p. 69

1.

a. $3x^4y^5y^2 \cdot 6x^6 = 18x^{10}y^7$	b. $(3x)^{-3} = \dfrac{1}{27x^3}$	c. $(3yz)^2 = 9y^2z^2$	d. $(b^{-2})^4 = \dfrac{1}{b^8}$
e. $\dfrac{8x^5}{28x^8} = \dfrac{2}{7x^3}$	f. $\dfrac{x^{-5}}{x^2} = \dfrac{1}{x^7}$	g. $\left(\dfrac{-2}{5y}\right)^2 = \dfrac{4}{25y^2}$	h. $\left(\dfrac{3s}{t^2}\right)^4 = \dfrac{81s^4}{t^8}$

Mixed Review 9, cont.

2\.

a. from point A with scale factor 1/3

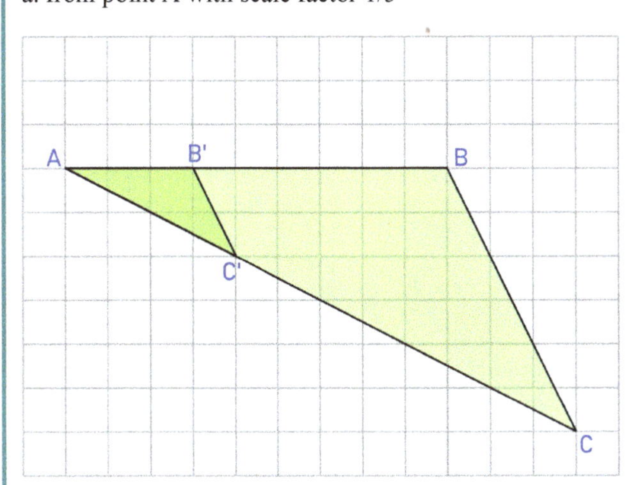

b. from point C with scale factor 1/2

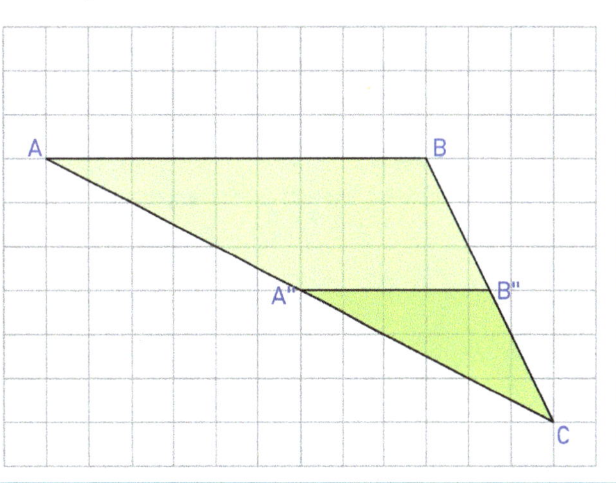

3\. The ratio of the volume of the tennis balls to the volume of the cube is

$$\frac{8 \cdot (4/3) \cdot \pi \cdot (3\text{ cm})^3}{(12\text{ cm})^3} = \frac{(32/3) \cdot \pi \cdot 27\text{ cm}^3}{12^3\text{ cm}^3} = \frac{32 \cdot \pi \cdot 27\text{ cm}^3}{3 \cdot 12^3\text{ cm}^3} = \frac{32 \cdot \pi \cdot 9}{3 \cdot 4 \cdot 3 \cdot 4 \cdot 3 \cdot 4} = \frac{1 \cdot \pi}{3 \cdot 2} = \frac{\pi}{6} \approx 52.4\%.$$

4\. The equation is $5 \cdot 5.95 + 5p = 53$. Here is the solution:

$$\begin{aligned} 5 \cdot 5.95 + 5p &= 53 \\ 29.75 + 5p &= 53 \\ 5p &= 23.25 \\ p &= 4.65 \end{aligned}$$

The unknown, discounted price was $4.65 per yard.

5\. a. Answers will vary. Check the student's answers. The first function should be of the form $y = mx$, or if using C for the cost and t for the time, of the form $C = mt$ for some constant m. Additionally, m can be a maximum of $50/8 = 6.25$ since the cost for eight hours has to be $50 or less. For example:

Function 1: $C = 6t$

Function 2:

time (hours)	0	1	2	3	4	5	6	7	8
Cost ($)	0	6	12	17	22	27	31	35	39

b. Answers will vary. Check the student's answers. Using the functions above, both functions give the same price for renting a surfboard for 2 hours ($12). For 6 hours, Function 2 gives the better deal, for $31 (versus $36).

Mixed Review 9, cont.

6. a. We can start with the equation $y = -2x + b$.
   To find b, we substitute −2 and 6 for x and y:
   $6 = -2(-2) + b$, from which $b = 6 - 4 = 2$.
   So, the equation is $y = -2x + 2$.

   b. We can start with the equation $y = (2/3)x + b$.
   To find b, we substitute 4 and −4 for x and y:
   $-4 = (2/3)4 + b$, from which $b = -4 - (2/3)4$
   $= -4 - 8/3 = -20/3$.
   So, the equation is $y = (2/3)x - 20/3$.

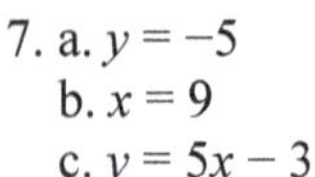

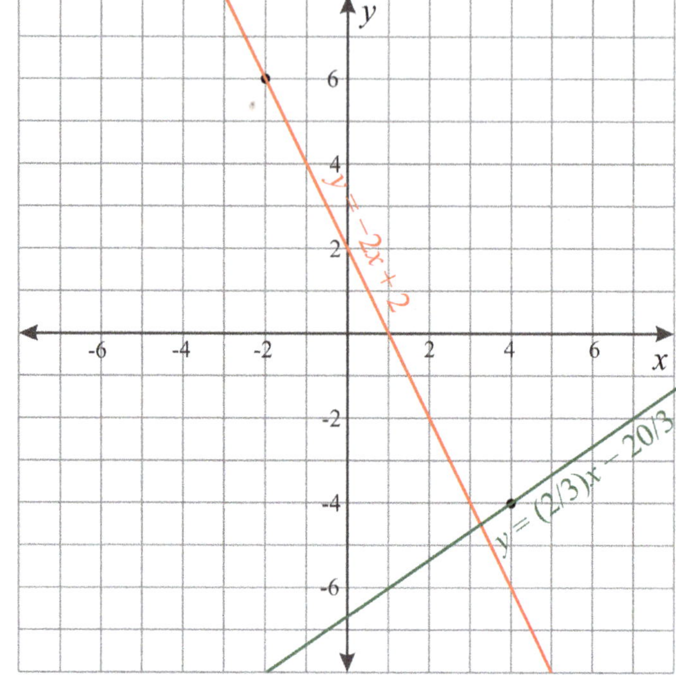

7. a. $y = -5$
   b. $x = 9$
   c. $y = 5x - 3$

8. a. 28.0 in b. 5,300 lb
   c. 71.6 m d. 38.0 kg
   e. 51.48 ft f. 11 cm

9. The volume of the square prism involved is found by simply multiplying the area of the bottom square by its height: 1 sq. mi · 1 inch. The first task is to convert one square mile to square inches.

   One mile = 5,280 ft = 5,280 ft · 12 in/ft = 63,360 inches.
   Then, a square mile is $(63{,}360 \text{ in})^2$, and the volume of our rainwater is then $1 \text{ in} \cdot (63{,}360 \text{ in})^2 = 4{,}014{,}489{,}600 \text{ in}^3$.

   Now, converting this to gallons, we get: $V = 4{,}014{,}489{,}600 \text{ in}^3 \cdot 1 \text{ gal}/(231 \text{ in}^3)$ = or in a simpler format, 4,014,489,600/231 gallons.

   If you calculate this intermediate result, keep a few more than four significant digits:
   $4{,}014{,}489{,}600/231 \approx 17{,}378{,}743$ gallons.

   And each of these gallons has 10^5 raindrops. So, we take the previous result and multiply it by 10^5:
   The number of raindrops $= 10^5 \cdot 17{,}378{,}743 \approx 1.738 \cdot 10^{12}$ drops.

Mixed Review 10, p. 72

1.

a.

$2 - \dfrac{x+1}{5}$	=	x	· 5
$10 - (x + 1)$	=	$5x$	
$10 - x - 1$	=	$5x$	
$9 - x$	=	$5x$	
9	=	$6x$	÷ 6
x	=	$3/2$	

b.

$4y + \dfrac{1-2y}{10}$	=	$3y$	· 10
$40y + (1 - 2y)$	=	$30y$	
$38y + 1$	=	$30y$	
$8y$	=	-1	÷ 8
y	=	$-1/8$	

c.

$\dfrac{x-10}{3}$	=	$\dfrac{3x+4}{5}$	· 15 (or cross-multiply)
$5(x - 10)$	=	$3(3x + 4)$	
$5x - 50$	=	$9x + 12$	
$-4x$	=	62	
x	=	$-31/2$	

d.

$-x + \dfrac{7x-3}{4}$	=	$\dfrac{x}{2} - 2$	· 8
$-8x + 2(7x - 3)$	=	$4x - 16$	
$-8x + 14x - 6$	=	$4x - 16$	
$6x - 6$	=	$4x - 16$	
$2x$	=	-10	
x	=	-5	

Mixed Review 10, cont.

2. a. $7.2 \cdot 10^{-5}$ b. $2.33 \cdot 10^{8}$ c. $4 \cdot 10^{-4}$ d. $3.09 \cdot 10^{8}$

3. a. > b. > c. <
   d. = e. < f. <

4. a. No solutions.
   b. One solution.
   c. One solution.
   d. An infinite number of solutions.

5. Of these, (2) and (3) are functions. In (1) the input 131 has two outputs. In (4), each input (each date) has millions of outputs, if you consider the population of the entire world.

6. a. Function 1 (90).
   b. Functions 2 and 3 are linear.
   c. Function 1. Its rate of change in that interval is $-50/4 = -25/2$.
   d. Function 1 is constant. Function 2 is increasing. Function 3 is decreasing.

7. No, it is not. It is not of the form $y = mx + b$ (or, using A and t, $A = ma + b$). Another way to see this is to calculate some values, and check whether the rate of change stays the same.

   We can easily see that as the t-values increase by ones, the A-values do not increase by the same amount each time. So, the function is not linear. In fact, this is an example of an *exponential function.*

t	A
0	5,000
1	5,300
2	5,618
3	5,955.08

8. a. has only one answer (see below), but the answer for (b) will vary. Check the student's answer. The output values should NOT always decrease or increase by the same amount.

a.

Input (x)	−5	−4	−3	−2	−1	0	1	2	3	4	5
Output (y)	−12	−9	−6	−3	0	3	6	9	12	15	18

b. For example:

Input (x)	−5	−4	−3	−2	−1	0	1	2	3	4	5
Output (y)	63	56	49	42	35	−2	5	6	−9	23	0

9. $V = 30 \text{ cm} \cdot 30 \text{ cm} \cdot 24 \text{ cm} / 3 = 7{,}200 \text{ cm}^3$

10. a. Company 1: $C = 42h + 1800$. Company 2: $C = 36h + 2400$.
    b. Company 1 will charge $2,430, whereas Company 2 will charge $2,940. <u>Company 1 is the better deal</u>.
    c. We will set the expressions for the costs equal, and solve that equation for h:

$$\begin{aligned} 42h + 1800 &= 36h + 2400 \\ 6h &= 600 \\ h &= 100 \end{aligned}$$

    For 100 hours of work, both companies will cost the same ($6,000).

Systems of Linear Equations Review, p. 75

1. a. An infinite number of solutions, because the second equation is two times the first, so they're equivalent equations.
   b. No solutions, because $7x - 2y$ cannot equal both 3 and −1.
   c. One solution. Considering these as lines, they have different slopes (−1 and 3/2).

2. a. (−1, 2) b. All points (x, y) that satisfy $2y - x = 1$ (or that satisfy $y = (1/2)x + 1/2$).

3. (b) has a single solution. $\begin{cases} y = (1/2)x + 6 \\ x + 6y = 0 \end{cases}$

Substituting $y = (1/2)x + 6$ to the second equation:

$$\begin{aligned} x + 6((1/2)x + 6) &= 0 \\ x + 3x + 36 &= 0 \\ 4x &= -36 \\ x &= -36/4 = -9 \end{aligned}$$

Now, we substitute $x = -9$ in the first equation: $y = (1/2)(-9) + 6 = -9/2 + 6 = -9/2 + 12/2 = 3/2$.
Solution: $(-9, 3/2)$.

4. $\begin{cases} x - 2y = 3 \\ 2x + y = -8 \end{cases}$

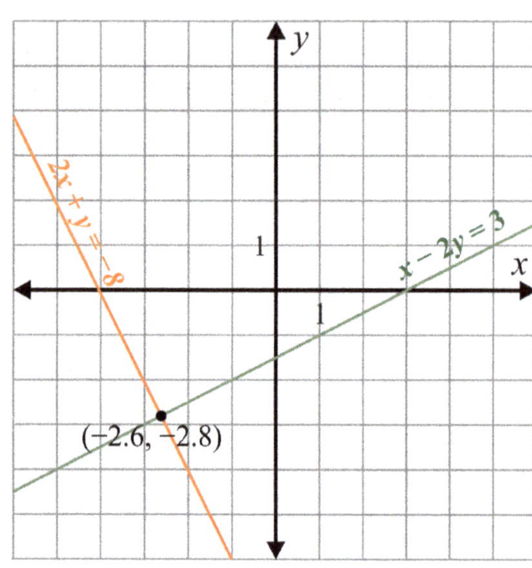

We can solve from the first equation that $x = 2y + 3$. Substituting $2y + 3$ for x in the second equation:

$$\begin{aligned} 2(2y + 3) + y &= -8 \\ 4y + 6 + y &= -8 \\ 5y + 6 &= -8 \\ 5y &= -14 \\ y &= -14/5 \end{aligned}$$

Now, we substitute $y = -14/5$ in the first equation:

$$\begin{aligned} x - 2(-14/5) &= 3 \\ x + 28/5 &= 3 \\ x &= 3 - 28/5 = 15/5 - 28/5 = -13/5 \end{aligned}$$

Solution: $(-13/5, -14/5)$ or $(-2.6, -2.8)$. This is verified by the graph.

5.

a. $\begin{cases} x = -(y - 5) \\ 2y = 12(1 - x) \end{cases}$

↓

$\begin{cases} x = -y + 5 \\ 2y = 12 - 12x \end{cases}$

↓

$\begin{cases} x + y = 5 & \vert \cdot (-12) \\ 12x + 2y = 12 \end{cases}$

↓

$$\begin{aligned} & -12x - 12y = -60 \\ + \quad & 12x + 2y = 12 \\ \hline & -10y = -48 \\ & y = 4.8 \end{aligned}$$

Substituting $y = 4.8$ in the first equation, we get:

$x = -(4.8 - 5)$

$x = -(-0.2) = 0.2$

Solution: $(0.2, 4.8)$ or $(1/5, 24/5)$

b. $\begin{cases} 4.5x - 3y = 0 & \vert \cdot 2 \\ 7x + 2y = -2 & \vert \cdot 3 \end{cases}$

↓

$$\begin{aligned} & 9x - 6y = 0 \\ + \quad & 21x + 6y = -6 \\ \hline & 30x = -6 \\ & x = -6/30 = -1/5 \end{aligned}$$

Substituting $x = -1/5$ (or -0.2 if you prefer) in the second equation, we get:

$$\begin{aligned} 7(-1/5) + 2y &= -2 \\ -7/5 + 2y &= -2 \\ 2y &= 7/5 - 2 \\ 2y &= -3/5 \\ y &= -3/10 \end{aligned}$$

Solution: $(-1/5, -3/10)$ or $(-0.2, -0.3)$

5.

c. $\begin{cases} 3x = -9(y + 1/3) \\ -2x - 6y = 2 \end{cases}$

$\downarrow$

$\begin{cases} 3x = -9y - 3 \\ -2x - 6y = 2 \end{cases}$

$\downarrow$

$\begin{cases} 3x + 9y = -3 & | \cdot 2 \\ -2x - 6y = 2 & | \cdot 3 \end{cases}$

$\downarrow$

$$\begin{array}{rrcl} & 6x + 18y & = & -6 \\ + & -6x - 18y & = & 6 \\ \hline & 0 & = & 0 \end{array}$$

Solution: all points (x, y) that satisfy the equation $-2x - 6y = 2$.

(The student can list some other, equivalent equation instead.)

d. $\begin{cases} 5x - 3(y + 2) = 0 \\ -5x + 6y = 7 \end{cases}$

$\downarrow$

$\begin{cases} 5x - 3y - 6 = 0 \\ -5x + 6y = 7 \end{cases}$

$\downarrow$

$$\begin{array}{rrcl} & 5x - 3y & = & 6 \\ + & -5x + 6y & = & 7 \\ \hline & 3y & = & 13 \\ & y & = & 13/3 \end{array}$$

Substituting $y = 13/3$ in the second equation, we get:

$$\begin{aligned} -5x + 6(13/3) &= 7 \\ -5x + 26 &= 7 \\ -5x &= -19 \\ x &= 19/5 \end{aligned}$$

Solution: (19/5, 13/3)

6. The error is in the highlighted portion. Instead, $y = 36/30 = 6/5$. The corrected solution is on the right.

Incorrect solution:

$\begin{cases} (1) \quad 6x + 3y = 3 \\ (2) \quad -2x + 9y = 11 & | \cdot 3 \end{cases}$

$\downarrow$

$$\begin{array}{r} 6x + 3y = 3 \\ -6x + 27y = 33 \\ \hline 30y = 36 \\ y = 5/6 \end{array}$$

Corrected solution:

$\begin{cases} (1) \quad 6x + 3y = 3 \\ (2) \quad -2x + 9y = 11 & | \cdot 3 \end{cases}$

$\downarrow$

$$\begin{array}{r} 6x + 3y = 3 \\ -6x + 27y = 33 \\ \hline 30y = 36 \\ y = 6/5 \end{array}$$

$\downarrow$

$$\begin{aligned} (1) \quad 6x + 3(6/5) &= 3 \\ 6x + 18/5 &= 3 \\ 6x &= 3 - 18/5 \\ 6x &= -3/5 \\ x &= -3/30 = -1/10 \end{aligned}$$

The solution (−1/10, 6/5) does fulfill the 2nd equation:

$$\begin{aligned} -2(-1/10) + 9(6/5) &\stackrel{?}{=} 11 \\ 1/5 + 54/5 &\stackrel{?}{=} 11 \\ 55/5 &= 11 \end{aligned}$$

7.

a.
$$\begin{cases} 7x + 6y = -1 \\ 11x + 2y = 3 \quad | \cdot (-3) \end{cases}$$
$$\downarrow$$
$$+\begin{cases} 7x + 6y = -1 \\ -33x - 6y = -9 \end{cases}$$
$$-26x = -10$$
$$x = 10/26 \approx 0.38462$$

Substituting $x = 0.38462$ in the first equation, we get:

$$\begin{aligned} 7(0.38462) + 6y &= -1 \\ 6y &= -1 - 7(0.38462) \\ y &= -0.61539 \end{aligned}$$

Solution: (0.385, −0.615)

b.
$$\begin{cases} 3.4x + 0.7y = 5 \quad | \cdot 2 \\ 0.5x - 0.2y = -2 \quad | \cdot 7 \end{cases}$$
$$\downarrow$$
$$+\begin{cases} 6.8x + 1.4y = 10 \\ 3.5x - 1.4y = -14 \end{cases}$$
$$10.3x = -4$$
$$x \approx -0.38835$$

Substituting $x = -0.38835$ in the first equation, we get:

$$\begin{aligned} 3.4(-0.38835) + 0.7y &= 5 \\ -1.32039 + 0.7y &= 5 \\ 0.7y &= 5 + 1.32039 \\ y &= 9.02913 \end{aligned}$$

Solution: (−0.388, 9.029)

8. Let b be the number of bikes, and t be the number of trikes. Then:

$$\begin{cases} b + t = 31 \\ 2b + 3t = 80 \end{cases}$$

From the top equation, we can solve that $b = 31 - t$. We will then substitute that in place of b in the second equation:

$$\begin{aligned} 2(31 - t) + 3t &= 80 \\ 62 - 2t + 3t &= 80 \\ t &= 18 \end{aligned}$$

Then, $b = 31 - 18 = 13$. There are 13 bikes and 18 trikes.

9. Let D be Denny's age, and S be Sam's age. Then:

$$\begin{cases} S + 4 = (2/3)(D + 4) \\ D = 2S - 14 \end{cases}$$

Substituting 2S − 14 for D in the first equation:

$$\begin{aligned} S + 4 &= (2/3)((2S - 14) + 4) \quad | \cdot 3 \\ 3S + 12 &= 2(2S - 10) \\ 3S + 12 &= 4S - 20 \\ -S + 12 &= -20 \\ -S &= -32 \\ S &= 32 \end{aligned}$$

Then, D = 2(32) − 14 = 50. Denny is 50 and Sammy is 32 years old.

10. Let t be the tens digit and u be the ones digit of the two-digit number. Then:

$$\begin{cases} t + u = 11 \\ 10t + u = 7u - 2 \end{cases}$$
$$\downarrow$$
$$\begin{cases} t + u = 11 \quad | \cdot 6 \\ 10t - 6u = -2 \end{cases}$$
$$\downarrow$$
$$+\begin{cases} 6t + 6u = 66 \\ 10t - 6u = -2 \end{cases}$$
$$16t = 64$$
$$t = 4$$

Then, $u = 11 - 4 = 7$. The number is 47.

11.

	weight	protein content	*protein amount*
peanut butter	w_1	25%	$0.25w_1$
protein powder	w_2	90%	$0.9w_2$
Mixture	210 g	30%	0.3(210) = 63 g

Equations: $\begin{cases} w_1 + w_2 = 210 \\ 0.25w_1 + 0.9w_2 = 63 \end{cases}$

Substituting $210 - w_2$ for w_1 in the second equation:

$$\begin{aligned} 0.25(210 - w_2) + 0.9w_2 &= 63 \\ 52.5 - 0.25w_2 + 0.9w_2 &= 63 \\ 0.65w_2 &= 10.5 \\ w_2 &\approx 16.154 \text{ g} \end{aligned}$$

Then, $w_1 = 210 - w_2 = 210\text{ g} - 16.154\text{ g} = 193.846$ g. The mixture contains 194 grams of peanut butter and 16 grams of protein powder.

12. Let v_w be the speed of the wind and v_c the speed of the crow.

	distance	*velocity*	*time*
Flying with a tailwind	400	$v_c + v_w$	48 sec
Flying with a headwind	400	$v_c - v_w$	72 sec

Using $d = vt$, the two equations we get are:

$$\begin{cases} 400 = 48(v_c + v_w) \\ 400 = 72(v_c - v_w) \end{cases}$$

$$\downarrow$$

$$\begin{cases} 400 = 48v_c + 48v_w & \div 4 \\ 400 = 72v_c - 72v_w & \div 4 \end{cases}$$

$$\downarrow$$

$$\begin{cases} 100 = 12v_c + 12v_w & \cdot 3 \\ 100 = 18v_c - 18v_w & \cdot 2 \end{cases}$$

$$\downarrow$$

$$\begin{aligned} & 300 = 36v_c + 36v_w \\ + \; & 200 = 36v_c - 36v_w \\ \hline & 500 = 72v_c \\ & v_c = 6.9\overline{4} \end{aligned}$$

Then, let's substitute $6.9\overline{4}$ for v_c in the first equation:

$$\begin{aligned} 400 &= 48(6.9\overline{4} + v_w) \\ 400 &= 333.\overline{3} + 48v_w \\ 48_w &= 66.\overline{6} \\ v_w &= 1.3\overline{8} \end{aligned}$$

The speed of the crow is about 6.9 m/s and the wind speed is about 1.4 m/s.

Systems of Linear Equations Test, p. 80

1. a. One solution.
   b. An infinite number of solutions.
   c. No solutions.

2. a. s = 3
   b. See the graph on the right.
   c. The lines are parallel and never intersect.

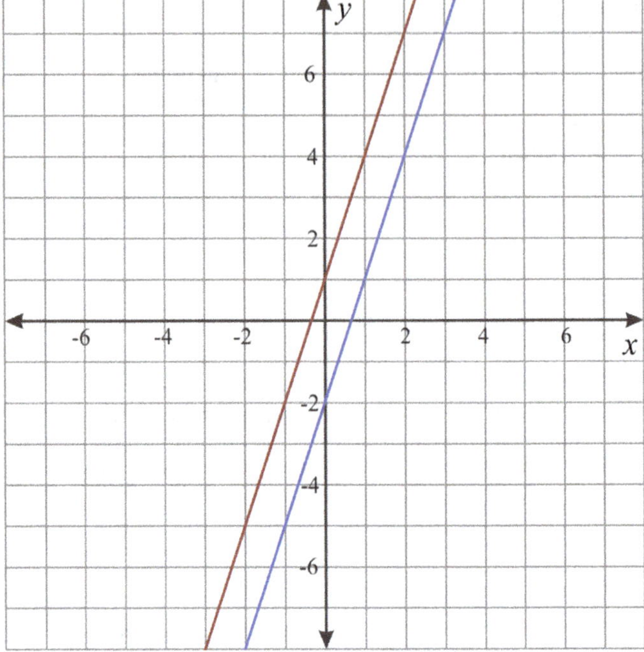

3.

a. $\begin{cases} x = 10 - y \\ 5y = 7(x - 2) \end{cases}$

$\downarrow$

$\begin{cases} x + y = 10 \\ 5y = 7x - 14 \end{cases}$

$\downarrow$

$\begin{cases} x + y = 10 \\ -7x + 5y = -14 \end{cases}$ $\Big|\cdot 7$

$\downarrow$

$$+\begin{cases} 7x + 7y = 70 \\ -7x + 5y = -14 \end{cases}$$

$$12y = 56$$

$$y = 56/12 = 14/3$$

Substituting $y = 14/3$ in the first equation, we get:

$x = 10 - 14/3$

$x = 30/3 - 14/3 = 16/3$

Solution: (16/3, 14/3)

b. $\begin{cases} 3x - y = 14 \\ 6x + 3y = -12 \end{cases}$ $\Big|\cdot 3$

$\downarrow$

$$+\begin{cases} 9x - 3y = 42 \\ 6x + 3y = -12 \end{cases}$$

$$15x = 30$$

$$x = 2$$

Substituting $x = 2$ in the first equation, we get:

$3(2) - y = 14$

$6 - y = 14$

$y = -8$

Solution: (2, −8)

4.

$$\begin{cases} x + 3y = -2 \\ 2x - 5y = -15 \end{cases}$$

Solving for x in the top equation, we get $x = -2 - 3y$.
Now substituting that in place of x in the second, we get:

$$\begin{aligned} 2(-2 - 3y) - 5y &= -15 \\ -4 - 6y - 5y &= -15 \\ -4 - 11y &= -15 \\ -11y &= -11 \\ y &= 1 \end{aligned}$$

Then, substituting 1 for y in the top equation, we get:

$$\begin{aligned} x + 3(1) &= -2 \\ x + 3 &= -2 \\ x &= -5 \end{aligned}$$

Solution: (−5, 1)

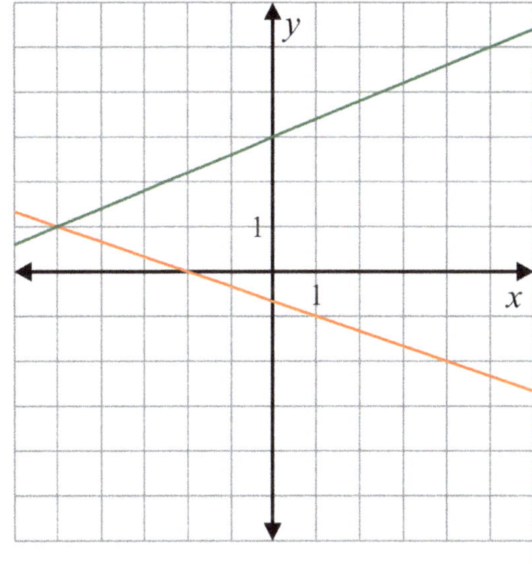

5.

$$\begin{cases} x + 6y = -1 \\ y = 2.2x + 0.8 \end{cases}$$

Substituting $2.2x + 0.8$ in place of y in the first equation, we get:

$$\begin{aligned} x + 6(2.2x + 0.8) &= -1 \\ x + 13.2x + 4.8 &= -1 \\ 14.2x + 4.8 &= -1 \\ 14.2x &= -5.8 \\ x &\approx -0.4084507 \end{aligned}$$

Then, substituting −0.4084507 for x in the top equation, we get:

$$\begin{aligned} -0.4084507 + 6y &= -1 \\ 6y &= -0.5915493 \\ y &\approx -0.09859155 \end{aligned}$$

Solution: (−0.408, −0.099)

6. Let A be Ann's age and D be Adam's age. Then:

$$\begin{cases} A + 3 = 2(D + 3) \\ A - 7 = 2.5(D - 7) \end{cases}$$

From the top equation, we get A = 2(D + 3) − 3 = 2D + 3. Substituting that for A in the bottom equation, we get:

$$\begin{aligned} 2D + 3 - 7 &= 2.5(D - 7) \\ 2D - 4 &= 2.5D - 17.5 \\ -4 &= 0.5D - 17.5 \\ 13.5 &= 0.5D \\ D &= 27 \end{aligned}$$

Then, A = 2D + 3 = 57.

Solution: Ann is 57 and Adam is 27.

7. Let x be the number of tables that seat 4 people, and y be the number of tables that seat 6. Then:

$$\begin{cases} x + y = 28 \\ 4x + 6y = 132 \end{cases}$$

We substitute $28 - x$ in place of y in the second equation, and get:

$$\begin{aligned} 4x + 6y &= 132 \\ 4x + 6(28 - x) &= 132 \\ 4x + 168 - 6x &= 132 \\ 168 - 2x &= 132 \\ -2x &= -36 \\ x &= 18 \end{aligned}$$

Then, $y = 28 - 18 = 10$. The restaurant has 18 tables that seat 4, and 10 tables that seat 6.

8. Let d be the distance that Train 1 travels until they meet and t be the amount of time since 3 PM when they meet. Then:

	distance	*velocity*	*time*
Train 1	d	90	t
Train 2	$70 - d$	105	t

Using the basic formula of $d = vt$ for both trains, we get the system of equations:

$$\begin{cases} d = 90t \\ 70 - d = 105t \end{cases}$$

We substitute $90t$ for d in the bottom equation:

$$\begin{aligned} 70 - 90t &= 105t \\ 70 &= 195t \\ t &= 70/195 = 14/39 \approx 0.358974 \end{aligned}$$

Then, $d = 90t = 90(14/39) \approx 32$ km. For Train 2, the distance is 70 km − 32 km = 38 km. The time 0.358974 hours is about 22 minutes.

So, the trains meet at 3:22 PM, and the second train has traveled about 38 km by that point.

Mixed Review 11, p. 83

1. Answers will vary; check the student's answer. For example:

First, reflect the trapezoid in the x-axis, so it becomes the trapezoid A'B'C'D'. Then, dilate it from point C' and with scale factor 1/2. Lastly, translate it 6.5 units to the right and one unit down.

Another possibility:

First, dilate the trapezoid from point D and with scale factor 1/2. Then, reflect it in the x-axis. Lastly, translate it six units to the right and two units down.

Mixed Review 11, cont.

2.

a.
$$\begin{aligned} s + 2 - 5s &= 3 - 8(s - 1) - 6s \\ 2 - 4s &= 3 - 8s + 8 - 6s \\ 2 - 4s &= 11 - 14s \\ 2 - 4s &= 11 - 14s \\ 10s &= 9 \\ s &= 9/10 \end{aligned}$$

b.
$$\begin{aligned} 6 + \frac{2x - 5}{3} &= x - 2 \quad \Big| \cdot 3 \\ 18 + (2x - 5) &= 3x - 6 \\ 13 + 2x &= 3x - 6 \\ -x &= -19 \\ x &= 19 \end{aligned}$$

c.
$$\begin{aligned} 0 &= \frac{x - 5}{3} + \frac{x + 5}{4} \quad \Big| \cdot 12 \\ 0 &= 4(x - 5) + 3(x + 5) \\ 0 &= 4x - 20 + 3x + 15 \\ 0 &= 7x - 5 \\ 7x &= 5 \\ x &= 5/7 \end{aligned}$$

d.
$$\begin{aligned} 10 &= y + \frac{2 - 3y}{5} \quad \Big| \cdot 5 \\ 50 &= 5y + (2 - 3y) \\ 50 &= 5y + 2 - 3y \\ 48 &= 2y \\ y &= 24 \end{aligned}$$

3. Answers will vary. Check the student's answers.

a. Here, the right side should have $2x$ and some constant that is not -5. For example: $2x - 5 = 2x + 2$. Or, $2x - 5 = 2(x - 3)$.

b. Here, the coefficient of the x term on the right side should be different from 2. For example: $2x - 5 = 3x + 1$, or $2x - 5 = x + 10 + 5x$.

c. Here, the right side should equal $2x - 5$. But we might make it look a little different, so that only after combining like terms does it equal $2x - 5$. For example: $2x - 5 = 2(x - 3) + 1$ or $2x - 5 = 6x - 4x - 7 + 2$.

4. a. $y = (3/4)x - 3$ b. $y = -(1/6)x + 3$ c. $y = -(1/2)x - 7/2$

5.

a.
$$\begin{cases} y = -2(x + 10) \\ 2(y - x) = 8 \end{cases}$$
$$\downarrow$$
$$\begin{cases} y = -2x - 20 \\ 2y - 2x = 8 \end{cases}$$
$$\downarrow$$
$$\begin{cases} 2x + y = -20 & \Big| \cdot (-2) \\ -2x + 2y = 8 \end{cases}$$
$$\downarrow$$
$$\begin{aligned} &\begin{cases} -4x - 2y = 40 \\ -2x + 2y = 8 \end{cases} \quad (+) \\ &-6x = 48 \\ &x = -8 \end{aligned}$$

Substituting $x = -8$ in the first equation, we get:
$$\begin{aligned} y &= -2(-8 + 10) \\ y &= -2(2) \\ y &= -4 \end{aligned}$$

Solution: $(-8, -4)$

b.
$$\begin{cases} 2x - 4y = -3 & \Big| \cdot 3 \\ 5x + 3y = 1 & \Big| \cdot 4 \end{cases}$$
$$\downarrow$$
$$\begin{aligned} &\begin{cases} 6x - 12y = -9 \\ 20x + 12y = 4 \end{cases} \quad (+) \\ &26x = -5 \\ &x = -5/26 \end{aligned}$$

Substituting $x = -5/26$ in the first equation, we get:
$$\begin{aligned} 2(-5/26) - 4y &= -3 \\ -10/26 - 4y &= -3 \\ -4y &= 10/26 - 3 \\ -4y &= 5/13 - 3 \\ -4y &= 5/13 - 39/13 \\ -4y &= -34/13 \\ y &= 34/52 = 17/26 \end{aligned}$$

Solution: $(-5/26, 17/26)$

6. a. Let d be the distance Charlie has run in 20 seconds. Then:

	distance	*velocity*	*time*
Rocky	$d + 8$	10 m/s	20 sec
Charlie	d	v	20 sec

Using $d = vt$, the two equations we get are:

$$\begin{cases} d + 8 = 10(20) \\ d = 20v \end{cases}$$

We will substitute $20v$ for d in the first equation:

$$\begin{aligned} 20v + 8 &= 200 \\ 20v &= 192 \\ v &= 192/20 = 9.6 \end{aligned}$$

So, Charlie's speed is 9.6 m/s.

b. We will use the same type of chart but just change Rocky's distance to $d + 2$:

	distance	*velocity*	*time*
Rocky	$d + 2$	10 m/s	20 sec
Charlie	d	v	20 sec

Using $d = vt$, the two equations we get are:

$$\begin{cases} d + 2 = 10(20) \\ d = 20v \end{cases}$$

We will substitute $20v$ for d in the first equation:

$$\begin{aligned} 20v + 2 &= 200 \\ 20v &= 198 \\ v &= 198/20 = 9.9 \end{aligned}$$

Charlie's speed would need to be 9.9 m/s so that he'd only be 2 m behind Rocky after 20 seconds.

7. We can look at the amount of copper and the amount of zinc in the mixture separately. The chart actually gives rise to three equations but we only need two.

	weight	*copper*	*amount of copper*	*zinc*	*amount of zinc*
Alloy 1	w_1	90%	$0.9w_1$	10%	$0.1w_1$
Alloy 2	w_2	65%	$0.65w_2$	35%	$0.35w_2$
Mixture	5000 kg	75%	3,750 kg	25%	1,250 kg

Equations: $\begin{cases} w_1 + w_2 = 5{,}000 \\ 0.9w_1 + 0.65w_2 = 3{,}750 \end{cases}$

(The third equation we could use comes from the amount of zinc: $0.1w_1 + 0.35w_2 = 1{,}250$.)
From the first equation, we can solve that $w_1 = 5{,}000 - w_2$. Substituting that into the second:

$$\begin{aligned} 0.9(5{,}000 - w_2) + 0.65w_2 &= 3{,}750 \\ 4{,}500 - 0.9w_2 + 0.65w_2 &= 3{,}750 \\ -0.25w_2 &= -750 \\ w_2 &= 3{,}000 \end{aligned}$$

So, the weight of the alloy 2 in the mixture is 3,000 kg and the weight of the alloy 1 is 2,000 kg.
Lastly, we might check that these values will fulfill the constraint for zinc:
0.1(2,000 kg) + 0.35(3,000 kg) = 200 kg + 1,050 kg = 1,250 kg. Yes, they do.

Mixed Review 11, cont.

8. a. In a minute, the elephant will cover $60 \cdot 6$ m = 360 m. So, his speed is 360 m/min.
   b. $d = 360t$
   c. Student graphs will vary because the student chooses the scaling for the vertical axis. For example:

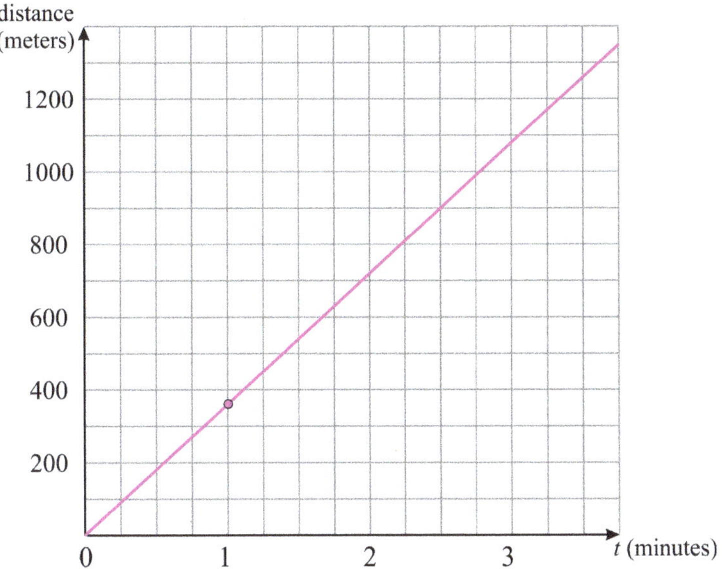

   d. The slope is 360 (or, 360 m/min).
   e. Elizabeth, because she goes at the speed of 400 meters per minute.
   f. Each minute, Elizabeth covers 40 m more than the elephant. So, in 3 minutes, she will cover 120 m more than the elephant.

Mixed Review 12, p. 87

1. $-2 \cdot 10^5 < -2 \cdot 10^{-5} < 2 \cdot 10^{-5} < 0.0005 < 2 \cdot 10^5$

2. Yes, they are. In triangle ABC, the third angle is 180° − 86° − 36° = 58°. This means its angles measure 86°, 58°, and 36°. Triangle DEF also has angles of 58° and 86°. So, based on the AA criterion for similar triangles, the triangles are similar.

   Alternatively, you could calculate the third angles for both triangles and note that both triangles have angles of 86°, 58°, and 36°, which means they are similar triangles.

3. a. Let C denote the cost and d the number of days. Then, $C = 55d + 10$.
   b. The rate of change is $55/day. It means that each additional day carries an additional cost of $55.
   c. The initial value is $10. It means that before you can even rent the snowboard, you have to pay a fixed $10 fee.

4.

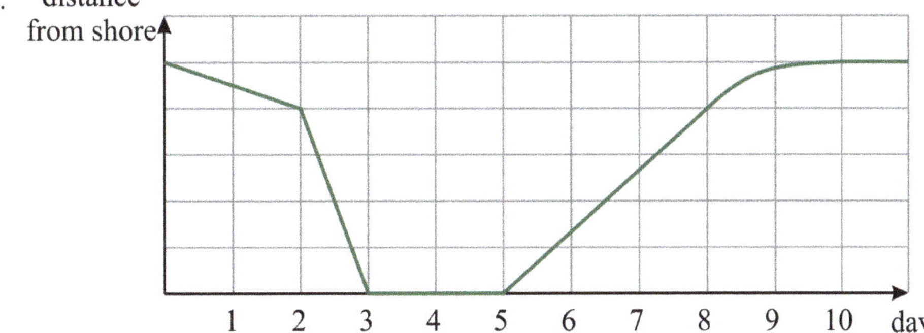

Mixed Review 12, cont.

5. Let l denote each of the legs of the isosceles triangle and b the base. Since the perimeter is 65, we can write the equation $2l + b = 65$. If the base would be 7 inches longer, the triangle would be equilateral, which means $b + 7 = l$. So, we have this system of equations:

$$\begin{cases} 2l + b = 65 \\ b + 7 = l \end{cases}$$

To solve it, we will substitute $b + 7$ for l in the top equation:

$$\begin{aligned} 2(b + 7) + b &= 65 \\ 2b + 14 + b &= 65 \\ 3b + 14 &= 65 \\ 3b &= 51 \\ b &= 17 \end{aligned}$$

Then, from the equation $b + 7 = l$ we can solve the $l = 17 + 7 = 24$.

The sides of the isosceles triangle measure 17, 24, and 24 units.

6. Angle BAD and the 52° angle are corresponding angles, which means ∠BAD = 52°. Then, ∠ BAC = ∠ BAD − ∠ CAD = 52° − 24° = 28°. Then, ∠ACD and ∠BAC are alternate interior angles, thus congruent. So, x = ∠BAC = 28°.

7. See the image on the right.

a. $y = (3/2)x - 2$, or in standard form, $3x - 2y = 4$.

b. The equation is of the form $y = -3x + b$. Substituting the point $(-4, 5)$ into this, we get $5 = -3(-4) + b$, from which $b = -7$. So, the equation of the line is $y = -3x - 7$, or in standard form, $3x + y = -7$.

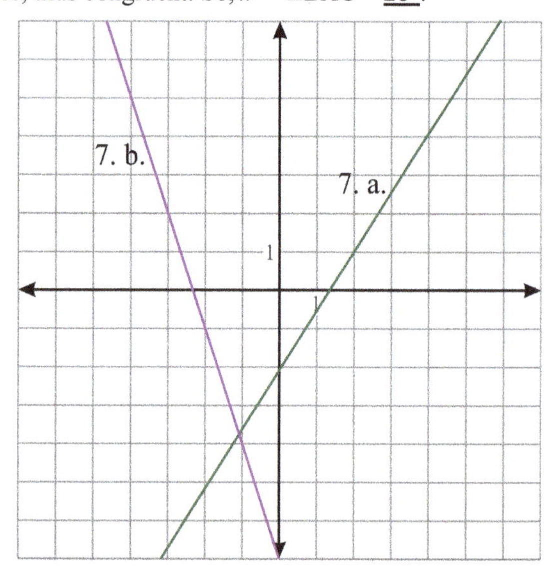

8.

a. Passes through (4, 1) and has slope 1/4 — **(iii)** $y = (1/4)x$

b. Passes through (−1/4, 1/4) and has slope 0 — **(iv)** $y = 1/4$

c. Passes through (1/4, 1/4) and has no slope — **(ii)** $x = 1/4$

d. Passes through (4, −1) and (−8, −4) — **(i)** $y = (1/4)x - 2$

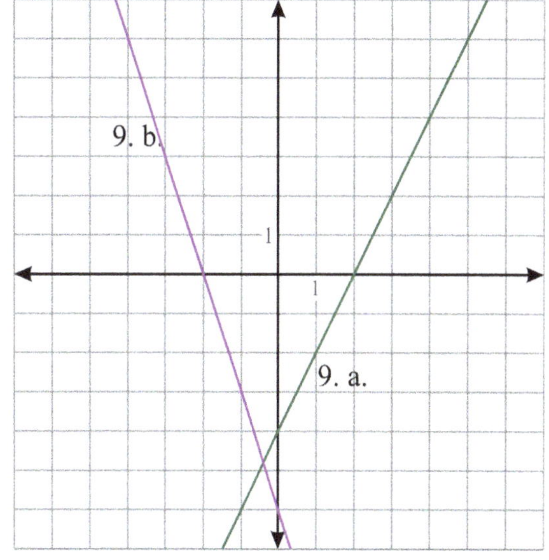

9. See the graphs on the right.
   a. The x-intercept is 2 and the y-intercept is −4.
   b. The x-intercept is −2 and the y-intercept is −6.

10. Let p be its price before the increases. Then,
    $1.03 \cdot 1.04 \cdot 1.09p = 268.55$, from which
    $p = 268.55/(1.03 \cdot 1.04 \cdot 1.09) = 230.00$.
    The item cost \$230 before the price increases.

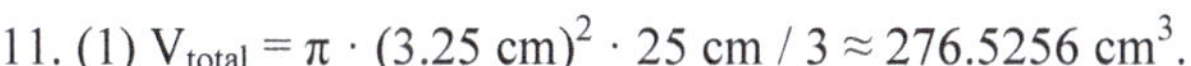

11. (1) $V_{total} = \pi \cdot (3.25\text{ cm})^2 \cdot 25\text{ cm} / 3 \approx 276.5256\text{ cm}^3$.

    (2) $V_{cut} = \pi \cdot (1.9\text{ cm})^2 \cdot 14.5\text{ cm} / 3 \approx 54.8156\text{ cm}^3$.

    (3) $V_{glass} = 276.5256\text{ cm}^3 - 54.8156\text{ cm}^3 \approx 220\text{ cm}^3$.

    Note that since most of the measurements were given to two significant digits, we give the final answer also to two significant digits.

Bivariate Data Review, p. 90

1. a. It is a negative linear association.

   b. There is at least one outlier: the country with about 89 years of life expectancy and a birth rate of 25 per 1000. Some other dots in the lower end of the life expectancy could perhaps be considered outliers, also, but without further statistical tools, we cannot say for sure.

   c. Answers will vary; check the student's trend line. For example:

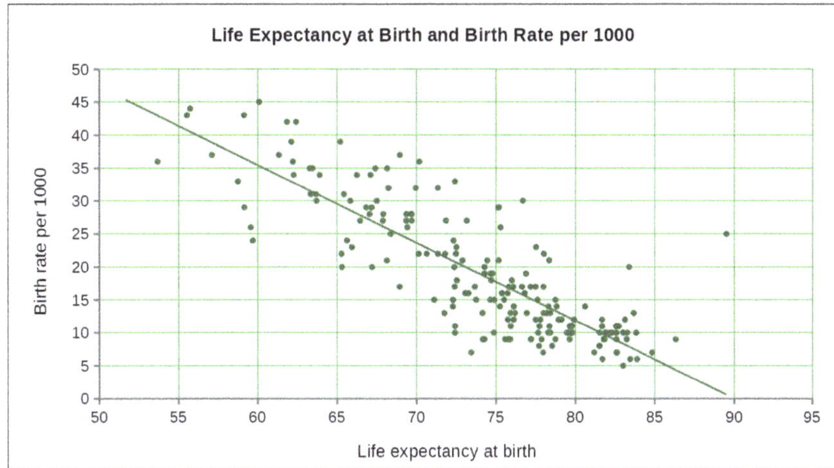

1. d. Answers will vary; check the student's trend line equation.

To find the equation, draw two points on the line, preferably points that are on the gridlines so that it will be easier to read their coordinates. Then figure out the slope using those two points: it is the ratio of (change in y-values/change in x-values).

In the case of the above graph, we could use for example (80, 12) and (70, 23.5) (estimated from the graph). The slope becomes $(12 - 23.5)/(80 - 70) = -11.5/10 = -1.15$.

Then, we figure out the y-intercept. The generic equation for a line is $y = mx + b$. We substitute into this the slope of -1.15 and one of the points, say (80, 12) to get the equation $12 = -1.15(80) + b$, from which $b = 104$.

So, the equation is $y = -1.15x + 104$.

For comparison, the equation, as calculated by a spreadsheet program, is $y = -1.181x + 106.3$. The equation we got, $y = -1.15x + 104$, is not exactly the same, but that is to be expected when we are estimating the coordinates of the points from the graph.

e. The slope of -1.15 means that each 1-year increase in life expectancy is associated with a decrease of 1.15 per 1000 in the birth rate.

f. We solve the equation $25 = -1.15x + 104$:

$$\begin{aligned} 25 &= -1.15x + 104 \\ 25 - 104 &= -1.15x \\ -79 &= -1.15x \\ x &= 79/1.15 \approx 69 \text{ years} \end{aligned}$$

g. Birth rate $= -1.15(55) + 104 \approx 41$ per 1000

h. We calculated the predicted life expectancy at birth when the birth rate is 25 per 1000 in question (f). It was 69 years, or more precisely, 68.7 years. The difference between 89.5 and that is 20.8 years. That's quite a bit!

2. a. We use the given equation of the line to calculate the coordinates for two points, and then we can draw the line.

The d-values for these two points can be any. Here, I chose 0 and 3,000 as being the beginning and end points of the horizontal axis.

First, if $d = 0$, then $C = 0.111(0) + 76.9 = 76.9$. Then, if $d = 3{,}000$, then $C = 0.111(3{,}000) + 76.9 = 409.9$. So, the two points to draw the line between are approximately (0, 77) and (3,000, 410).

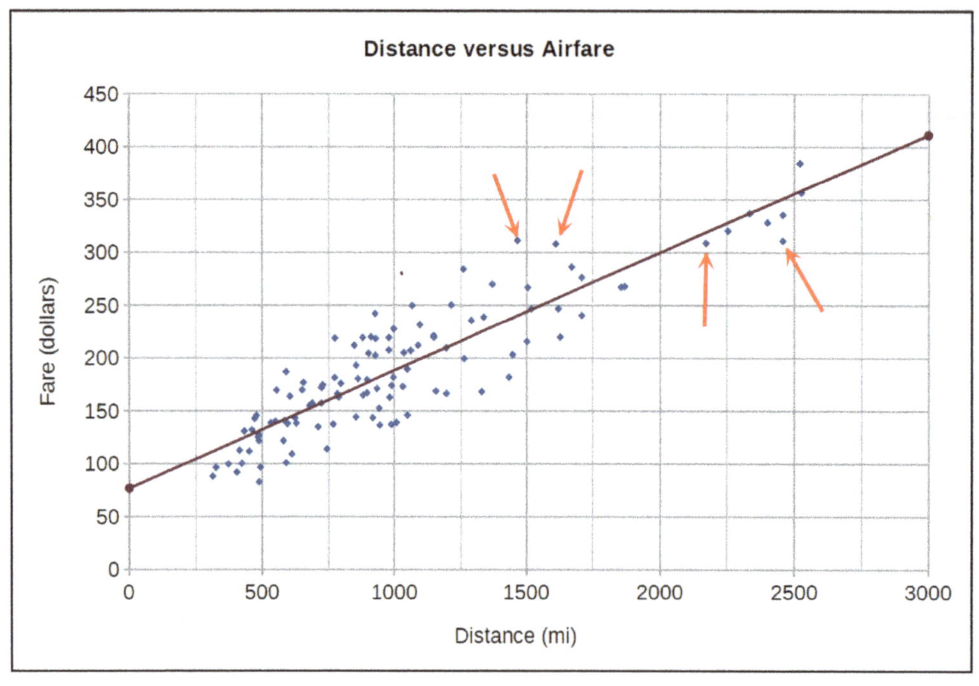

2. b. Those four points are signaled by the arrows in the image above. The distances are about 1,450, 1,600, 2,200, and 2,450 miles.

c. When the cost is \$310, we set C to be 310 in the equation $C = 0.111d + 76.9$ and solve for d:

$$310 = 0.111d + 76.9$$
$$310 - 76.9 = 0.111d$$
$$233.1 = 0.111d$$
$$d = 233.1/0.111 = 2{,}100$$

The equation predicts a distance of 2,100 miles.

d. When d is 800 miles, $C = 0.111(800) + 76.9 =$ \$165.70.

e. We use the slope. The slope of 0.111 signifies that each 1-mile increment in distance is associated with a \$0.111-increase in the airfare. This means that a 100-mile increase in distance is associated with a 100(0.111) = \$11.10 increase in airfare.

f. When the cost is \$300, the distance is 2,009.91 miles (solve the equation $300 = 0.111d + 76.9$ for d).

When the cost is \$250, the distance is 1,559.46 miles. The difference is 450.45 miles, or about 450 miles.

3. a.

	soccer	baseball	basketball	Total
Asia	17	1	5	23
Europe	85	7	60	152
Middle East	40	3	10	53
North America	11	45	26	82
Total	153	56	101	310

	soccer	baseball	basketball	Total
Asia	74%	4%	22%	100%
Europe	56%	5%	39%	100%
Middle East	75%	6%	19%	100%
North America	13%	55%	32%	100%

b. Answers will vary; check the student's answer. For example: Soccer is the most popular favorite sport everywhere except among the employees in North America, who prefer baseball. Baseball is not very popular outside of North American employees. Basketball is in the middle; about 1/3 of the North American respondents, about 2/5 of the European ones, and about 1/5 of the Middle Eastern and Asian ones consider it their favorite.

c. Yes, there is an association. The relative frequencies vary a lot based on the location. For example, soccer is the favorite of 75% of the respondents in Middle East, but only of 13% of those in North America. And baseball is not the preferred sport of many outside of North America.

d. $85/153 = 55.\overline{5}$. There is a 55.6% chance that that person is from Europe.

e. $10/101 \approx 0.0990$. There is a 9.9% chance that that person is from the Middle East.

f. $5/23 \approx 0.2174$. There is a 21.7% chance that that person prefers basketball.

4. a. 9
b. 2
c. 24

Bivariate Data Test, p. 93

1. a. Yes. There is a negative linear association between the variables.
   b. (24, 31.3)
   c. Student graphs will vary; check the student's graph. For one example, see the graph below. We're not looking for anything super accurate; the student's line should go through somewhere in the middle, without leaving many more points below the line than above it, or vice versa.

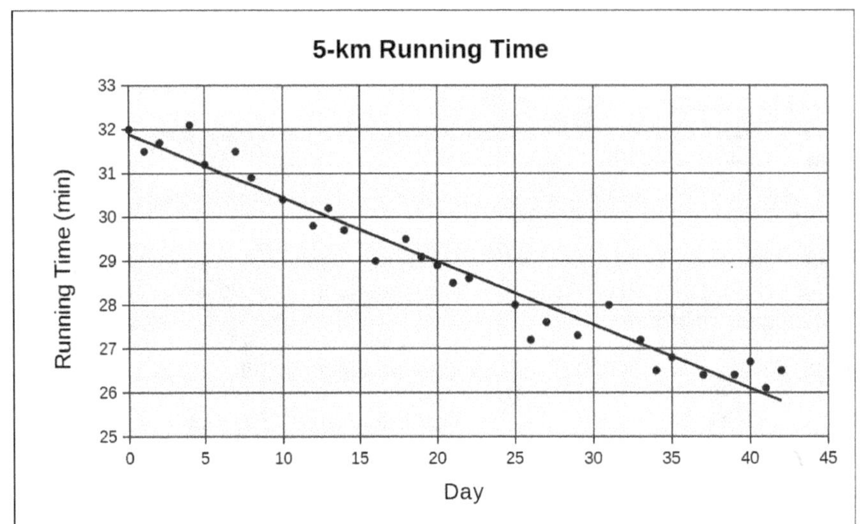

   d. Answers will vary; check the student's equation. For the line here, we can see it goes through (0, 31.9) approximately and (20, 29). From those two points, we can calculate the slope as $(29 - 31.9)/20 = -0.145$. The y-intercept is 31.9. So, the equation is $y = -0.145x + 31.9$, or, if using d for day and t for running time, $t = -0.145d + 31.9$. For comparison, the equation calculated by a spreadsheet program (for the line in this graph) is approximately $y = -0.144x + 31.887$.

   e. For day 24, the equation predicts a running time of $t = -0.145(24) + 31.9 = 28.42$ minutes.

   f. The slope of -0.145 signifies that for each additional day, the running time tends to decrease by 0.145 minutes.

   g. The y-intercept (or t-intercept) is 31.9 minutes. It signifies that at the start of the program, at zero days, the model predicts a running time of 31.9 minutes.

2. a. The slope of 1.34 signifies that each additional inch in diameter is associated with a \$1.34-price increase.

   b. No, we cannot. The equation predicts a negative price for a pizza with 1-inch diameter. (In fact, a pizza with 4.44-inch diameter would have a price of \$0, and smaller ones would have a negative price.)

   c. Price = $1.34(15) - 5.95 = \$14.15$.

   d. We solve the equation $15 = 1.34x - 5.95$ for x:

$$\begin{aligned} 15 &= 1.34x - 5.95 \\ 20.95 &= 1.34x \\ x &\approx 15.63 \end{aligned}$$

   The equation predicts a size of 15.6 inches for a pizza that costs \$15.

3. a. Yes. Since the total number of males and females is similar, if there wasn't an association, the numbers of students taking any of the languages, or no language, would also be similar. But instead, we see that twice as many females as males took Spanish, and more than twice as many females as males took French.
   b. $9/13 \approx 69\%$
   c. $117/145 \approx 81\%$

4. a. The table "Has two or more siblings" does not have any association between the variables.

   b. This can be seen by checking the relative frequencies. They could be calculated as percentages, however, in this case it is not even necessary to do that. It is easy to notice that in the sibling situation, for each age group, those who said "Yes" are about half of those who said "No" (which means the relative frequencies for "Yes" would be close to 33% in each age group, and for "No", about 67%). This table is the only one with that kind of situation.

Mixed Review 13, p. 96

1. See the image on the right. Angle 1 is supplementary to the 97° angle, so it is 83°. In triangle CBA, angle 2 is 180° − 35° − 83° = 62°. Angle 3 and angle 2 are vertical angles, so angle 3 is also 62°. In triangle AED, angle x is 180° − 62° − 46° = 72°.

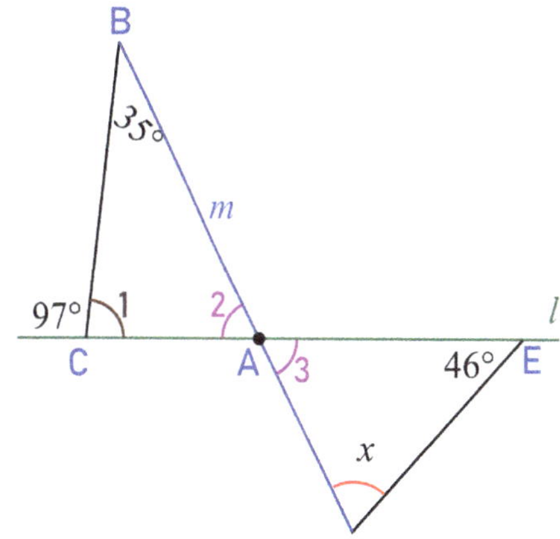

2. a. 8 b. 6 c. 9/4
   d. 0.4 e. 0.1 f. 1.1

3.

a.	b.	c.
$18 + t^3 = 134$	$5x^2 = 220$	$109 - v^3 = 11$
$t^3 = 116$	$x^2 = 44$	$109 - 11 = v^3$
$t = \sqrt[3]{116} \approx 4.877$	$x = \sqrt{44} \approx 6.633$	$98 = v^3$
	or $x = -\sqrt{44} \approx -6.633$	$v = \sqrt[3]{98} \approx 4.610$

4. a. Correct.
   b. Not correct. $\sqrt{121} = 11$ so it is a whole number.
   c. Correct.
   d. Not correct. Unending repeating decimals are rational; 0.19191919... is a rational number. As a fraction, it is 19/99.

5. Let s be the unknown side. The diagonal divides the rectangle into two right triangles. Applying the Pythagorean Theorem to one of them (on the right), we get:

$$
\begin{aligned}
2.80^2 + s^2 &= 5.5^2 \\
7.84 + s^2 &= 30.25 \\
s^2 &= 22.41 \\
x &= \sqrt{22.41} \approx 4.7
\end{aligned}
$$

The other side of the rectangle is 4.7 m long.

6. We need to first find the length marked with y (see the image on the right). After that, we can use the Pythagorean Theorem to find x.

To find y, we apply the Pythagorean Theorem in the triangle ABC:

$$
\begin{aligned}
0.6^2 + 1.6^2 &= y^2 \\
0.36 + 2.56 &= y^2 \\
y^2 &= 2.92 \\
y &= \sqrt{2.92}
\end{aligned}
$$

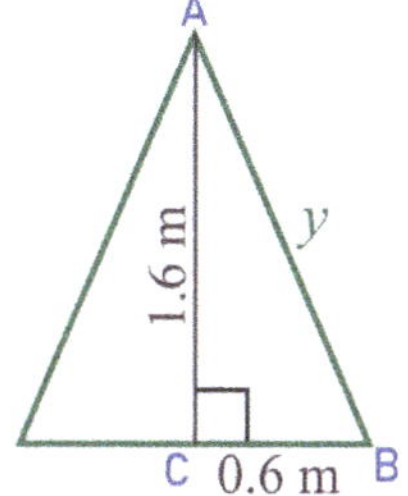

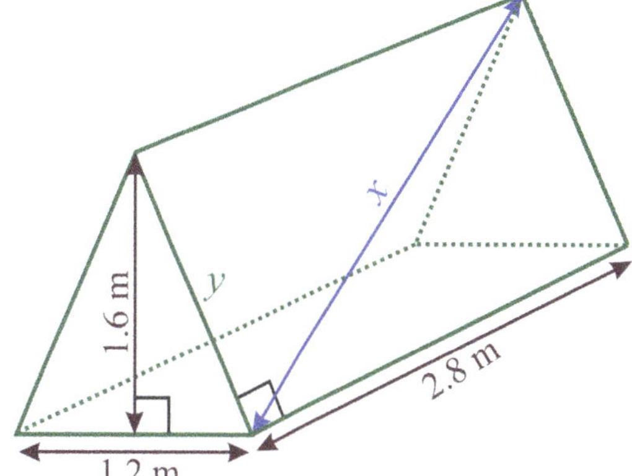

Then, we use the Pythagorean Theorem in the triangle that is part of the tent's side, with sides y, x, and 2.8 meters:

$$
\begin{aligned}
y^2 + 2.8^2 &= x^2 \\
(\sqrt{2.92})^2 + 2.8^2 &= x^2 \\
2.92 + 7.84 &= x^2 \\
y^2 &= 10.76 \\
y &= \sqrt{10.76} \approx 3.28
\end{aligned}
$$

The diagonal marked with x is 3.3 m long.

7. Answers will vary since the description did not give actual amounts of water.
The first portion should be linear and decreasing. The portion for September should be increasing in a nonlinear manner, so that it rises fast at first and then slows down. The portions for October and January should be horizontal lines. The portion for November-December should be slowly increasing and linear. For example:

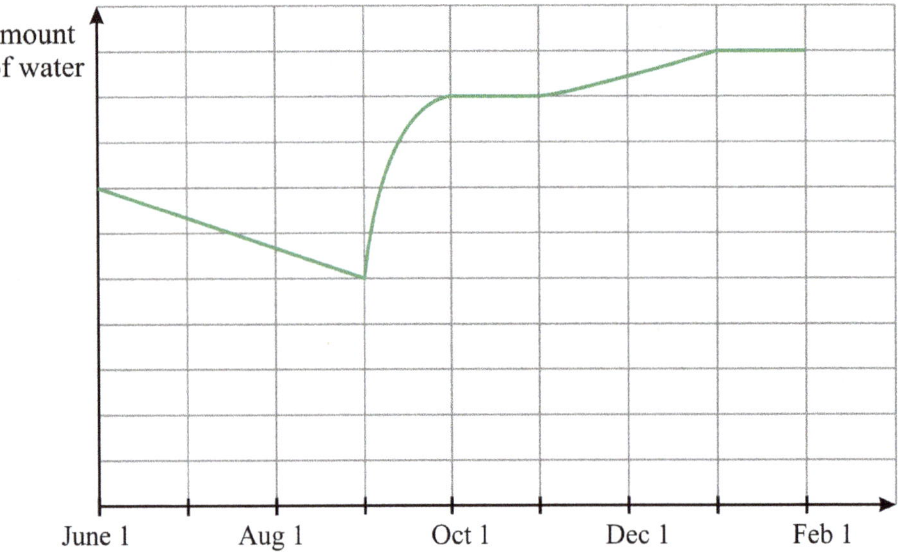

8. a. $y = (2/3)x - 2$

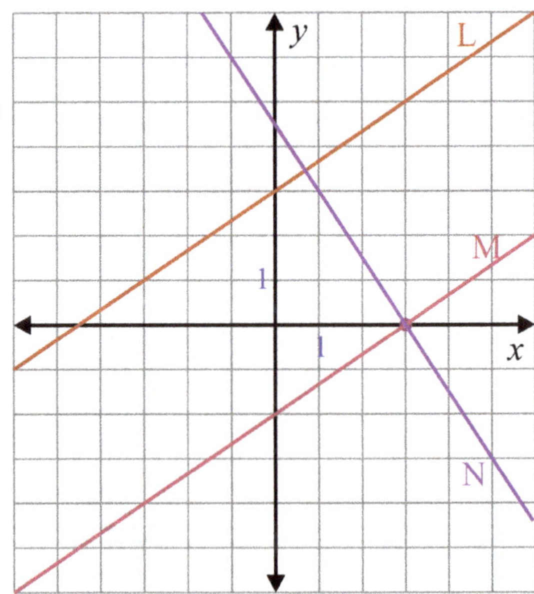

b. The slope of line N is $-3/2$. Its equation is of the form $y = -(3/2)x + b$. To find b, we will substitute (3, 0) into the equation and solve for b:

$0 = -(3/2)(3) + b$

$0 = -9/2 + b$

$b = 9/2$

So, the equation is $\underline{y = -(3/2)x + 9/2}$.

9.

a. $\begin{cases} 5x - 7y = 3 \\ -6x + 8y = 1 \end{cases}$ $\begin{vmatrix} \cdot 6 \\ \cdot 5 \end{vmatrix}$

$\downarrow$

$$\begin{array}{r} \begin{cases} 30x - 42y = 18 \\ -30x + 40y = 5 \end{cases} \\ \hline -2y = 23 \\ y = -23/2 \end{array}$$

Substituting $y = -23/2$ in the second equation, we get:

$$\begin{aligned} -6x + 8(-23/2) &= 1 \\ -6x - 92 &= 1 \\ -6x &= 93 \\ x &= -93/6 = -31/2 \end{aligned}$$

Solution: $(-31/2, -23/2)$ or $(-15.5, -11.5)$

b. $\begin{cases} 3x = -2(y + 1) \\ -x - 3y = 0 \end{cases}$

$\downarrow$

$\begin{cases} 3x = -2y - 2 \\ -x - 3y = 0 \end{cases}$

$\downarrow$

$\begin{cases} 3x + 2y = -2 \\ -x - 3y = 0 \end{cases}$ $| \cdot 3$

$\downarrow$

$$\begin{array}{r} \begin{cases} 3x + 2y = -2 \\ -3x - 9y = 0 \end{cases} \\ \hline -7y = -2 \\ y = 2/7 \end{array}$$

Substituting $y = 2/7$ in the second equation, we get:

$$\begin{aligned} -x - 3(2/7) &= 0 \\ -x - 6/7 &= 0 \\ x &= -6/7 \end{aligned}$$

Solution: $(-6/7, 2/7)$

10. Let x be the number of chickens and y be the number of cows. Then:

$$\begin{cases} x + y = 42 \\ 2x + 4y = 100 \end{cases}$$

Solving for y in the first equation, we get $y = 42 - x$. Substituting that in place of y in the second, we get:

$$\begin{aligned} 2x + 4(42 - x) &= 100 \\ 2x + 168 - 4x &= 100 \\ 168 - 2x &= 100 \\ -2x &= -68 \\ x &= 34 \end{aligned}$$

Then, $y = 42 - 34 = 8$. There are 34 chickens and 8 cows.

11. a. There is a positive linear association. We do not see any outliers or clusters.
b. Answers will vary; check the student's line. For example:

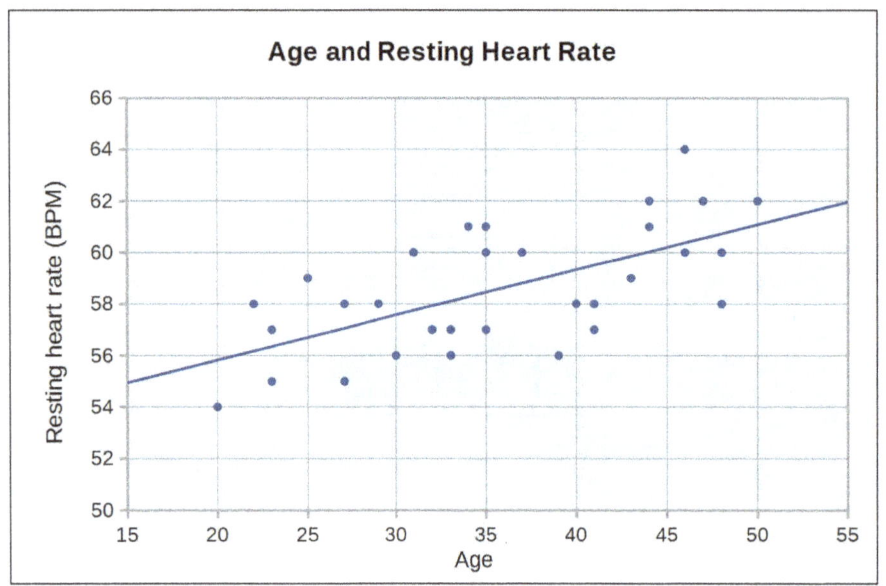

c. Answers will vary; check the student's equation. The equation for the line in the graph above is approximately $y = 0.18x + 52$.

d. Answers will vary; check the student's answer. Using the equation above, it increases by 0.18 beats per minute.

e. Answers will vary; check the student's answer. Using the equation above, if $x = 30$, we get the heart rate as $y = 0.18(30) + 52 = 57.4$ beats per minute.

f. For a newborn, the equation predicts a heart rate of $0.18(0) + 52 = 52$ beats per minute.
Since in reality, a newborn heart rate is 120 to 160 bpm, we cannot use this equation to extrapolate much (if any) past the range of the original data.

12. a. (i) and (iii)
b. See the chart on the right.
c. Yes. When you compare the percentages in the two columns, they are definitely different. The majority of female clients have haircuts either two or three months apart, whereas half of the male clients have a haircut every month. The males have haircuts more often than females, and that makes sense because shorter hairstyles need a haircut more often.

Relative frequencies:

	Male	Female
every month	50%	12%
every two months	28%	30%
every three months	15%	37%
every four months	7%	20%
TOTALS	100%	99%

Mixed Review 14, p. 101

1. The bottom row of the table gives the coordinates after all the transformations.

Original	A(5, −1)	B(3, −4)	C(1, −3)
After reflection	A'(5, 1)	B'(3, 4)	C'(1, 3)
After dilation	A"(10, 2)	B"(6, 8)	C"(2, 6)
After rotation	A'''(2, −10)	B'''(8, −6)	C'''(6, −2)

2. a. One solution.
   b. $10s - 6 = 10s$ or $2s - 6 = 2s$.

3.

a. $\begin{cases} 3x - y = 14 \\ 6x + 3y = -12 \end{cases}$

We can solve for y from the top equation: $y = 3x - 14$. Substituting that for y in the second equation, we get:

$$\begin{aligned} 6x + 3(3x - 14) &= -12 \\ 6x + 9x - 42 &= -12 \\ 15x - 42 &= -12 \\ 15x &= 30 \\ x &= 2 \end{aligned}$$

Then, substituting 2 for x in the top equation, we get:

$$\begin{aligned} 3(2) - y &= 14 \\ -y &= 14 - 6 = 8 \\ y &= -8 \end{aligned}$$

Solution: (2, −8)

b. $\begin{cases} 3x - 8(y - 2) = 0 \\ 2x = 8y + 1 \end{cases}$

↓

$\begin{cases} 3x - 8y + 16 = 0 \\ 2x = 8y + 1 \end{cases}$

↓

$\begin{cases} 3x - 8y = -16 \\ 2x - 8y = 1 \quad | \cdot (-1) \end{cases}$

↓

$$\begin{aligned} 3x - 8y &= -16 \\ + \quad -2x + 8y &= -1 \\ \hline x &= -17 \end{aligned}$$

Substituting $x = -17$ in the second equation, we get:

$$\begin{aligned} 2(-17) &= 8y + 1 \\ -34 &= 8y + 1 \\ 8y &= -35 \\ y &= -35/8 \end{aligned}$$

Solution: (−17, −35/8)

4. a. Not correct. Corrected form: $\sqrt{121}$ is rational because it is a square root of a perfect square (it equals 11).
   b. Not correct. Corrected form: 0.831831831 is rational because it is a terminating decimal. Or, because it is the fraction 831,831,831/1,000,000,000.
   c. Correct.
   d. Correct.
   e. Not correct. Corrected form: $\sqrt{63}/4$ is irrational because $\sqrt{63}$ is irrational, 4 is rational, and an irrational number divided by a rational number is irrational.

5. Let V be the volume of the 3% milk in this mixture. The chart looks like this:

	volume (ml)	*butterfat percentage*	*butterfat amount*
3% milk	V	3	0.03V
15% cream	1000 − V	15	0.15(1000 − V)
Mixture	1000	12	120

From the last column, we can write the equation $0.03V + 0.15(1000 - V) = 120$. Here is its solution:

$$\begin{aligned} 0.03V + 0.15(1000 - V) &= 120 \\ 0.03V + 150 - 0.15V &= 120 \\ 150 - 0.12V &= 120 \\ -0.12V &= -30 \\ V &= 250 \end{aligned}$$

You should mix 250 mL of the 3% milk and 750 mL of the 15% cream in order to get 1 liter of a mixture that has 12% butterfat.

6. In the spot marked below the variable x gets switched to y. The corrected solution is on the right.

(1) (2) $\begin{cases} 3x - 7y = 4 \\ -5x + y = -4 \end{cases} \Big| \cdot 7$

↓

$\begin{cases} 3x - 7y = 4 \\ -35x + 7y = -28 \end{cases}$

$-32y = -24$ ← Here is where the x and y got switched.

$y = 3/4$

↓

(1) $3x - 7(3/4) = 4$

$3x - 21/4 = 4$

$3x = 4 + 21/4 = 37/4$

$3x = 37/4$

$x = 37/12$

(1) (2) $\begin{cases} 3x - 7y = 4 \\ -5x + y = -4 \end{cases} \Big| \cdot 7$

↓

$\begin{cases} 3x - 7y = 4 \\ -35x + 7y = -28 \end{cases}$

$-32x = -24$

$x = 3/4$

↓

(1) $3(3/4) - 7y = 4$

$9/4 - 7y = 4$

$-7y = 4 - 9/4$

$-7y = 7/4$

$y = -1/4$

Solution: $(3/4, -1/4)$

7. We ignore the negative roots since these are lengths of sides of triangles.

a.	b.
$11.2^2 + 7.6^2 = s^2$	$(\sqrt{26})^2 + x^2 = 7^2$
$11.2^2 + 7.6^2 = s^2$	$26 + x^2 = 49$
$183.2 = s^2$	$x^2 = 23$
$s = \sqrt{183.2} \approx 13.5$	$x = \sqrt{23}$

8. Let h be the height of the triangle. We apply the Pythagorean Theorem in the triangle BCD, in order to find the height of the triangle:

$8.25^2 + h^2 = 16.5^2$

$h^2 = 16.5^2 - 8.25^2$

$h^2 = 204.1875$

$h = \sqrt{204.1875} \approx 14.289$ in

Then, the area is A = 16.5 in (14.289 in) / 2 ≈ 118 square inches.

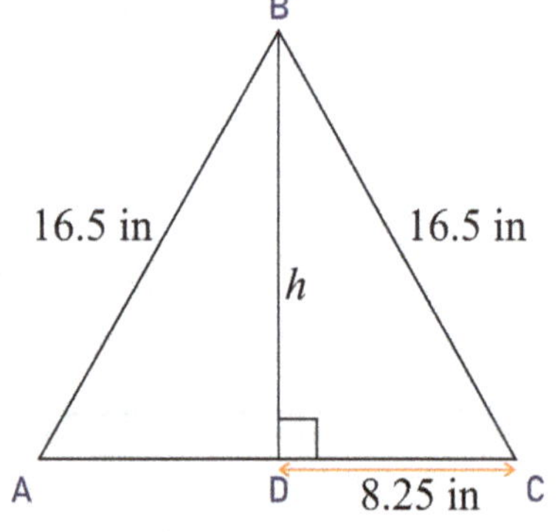

9. a. See the table on the right.

b. Yes. If there wasn't, we would see similar numbers for each hobby for boys and girls. But for example for arts and crafts, there are over twice as many girls than boys. Or, for reading, there are about 1.5 times as many girls as boys.

Favorite hobbies of 2nd graders

	Boys	Girls	Total
Sports	26	21	47
Music	12	13	25
Reading	13	20	33
Arts & Crafts	5	13	18
Video games	16	7	23
Cooking	2	5	7
Photography	6	5	11
Total	80	84	164

10. $V = \pi r^2 h$

$\frac{V}{\pi h} = r^2$

$r = \sqrt{\frac{V}{\pi h}}$

Since this has to do with the radius of a circle, we discard the negative root.

11. a. b. Answers will vary. Check the student's answer. Here are two examples:

(1)

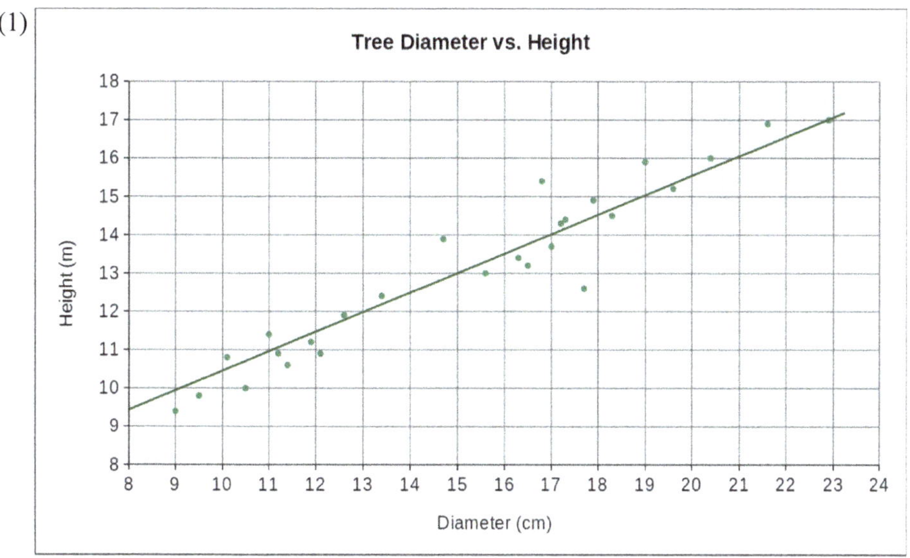

Approximate equation: The line goes through the points (8, 9.5) and (15, 13), from which the slope is 3.5/7 = 0.5. The equation is $y = 0.5x + b$. Substituting (15, 13), we get $13 = 0.5(15) + b$, from which $b = 5.5$. So, the equation is $y = 0.5x + 5.5$.

(2)

Approximate equation: $y = 0.55x + 4.77$ (calculated by a spreadsheet program).

c. The slope means that each additional centimeter in tree diameter is associated with a 0.5-meter (or 0.55-meter) increase in height.

d. It signifies that the model predicts that a tree with a diameter of zero centimeters is 5.5 m (4.77 m) high. This clearly doesn't make sense, so, we cannot extrapolate this model that much backwards.

e. Student answers will vary since the equations will vary. If the tree is 16.5 m tall, we will calculate the diameter using the equation $y = 0.5x + 5.5$:

$$\begin{aligned} 16.5 &= 0.5x + 5.5 \\ 0.5x &= 11 \\ x &= 22 \end{aligned}$$

The equation predicts that the diameter of the tree is 22 cm .

End-of-Year Test, p. 105

Use your judgment in grading. You can give points or partial points for partial answers.

Question #	Max. points	Student score
Exponents and Scientific Notation		
1	8 points	
2	9 points	
3	4 points	
4	2 points	
5	2 points	
	subtotal	/ 25
Irrational Numbers		
6	5 points	
7	5 points	
8	3 points	
9	2 points	
	subtotal	/ 15
Geometry		
10	3 points	
11	2 points	
12	3 points	
13	2 points	
14a	3 points	
14b	3 points	
15	3 points	
16	3 points	
	subtotal	/ 22
Linear Equations		
17	4 points	
18	4 points	
19	6 points	
20	2 points	
21	2 points	
22	3 points	
	subtotal	/21
Functions		
23	2 points	
24a	1 point	
24b	2 points	
24c	2 points	

Question #	Max. points	Student score
Functions		
25a	1 point	
25b	1 point	
25c	1 point	
25d	1 point	
25e	1 point	
26a	2 points	
26b	1 point	
26c	1 point	
26d	1 point	
	subtotal	/17
Graphing Linear Equations		
27a	1 point	
27b	1 point	
27c	2 points	
28	3 points	
29	3 points	
30	3 points	
	subtotal	/13
The Pythagorean Theorem		
31	4 points	
32	3 points	
33	3 points	
	subtotal	/10
Systems of Linear Equations		
34	6 points	
35	3 points	
36	3 points	
37	3 points	
	subtotal	/15
Bivariate Data		
38	3 points	
39	3 points	
40	3 points	
41	5 points	
	subtotal	/14
	TOTAL	/152

Exponents and Scientific Notation

1. a. −16 b. 16 c. 1/49 d. 36
   e. 0.031 f. 110,000 g. −8/27 h. 64

2.

a. $-8s^3$	b. $144x^2$	c. y^{15}
d. $-6x^8$	e. $\dfrac{1}{y^6}$	f. $\dfrac{1}{4v^2}$
g. $\dfrac{49x^2}{9y^2}$	h. $\dfrac{-x^3}{125}$	i. $\dfrac{81b^4}{c^{20}}$

3. a. $1.93 \cdot 10^8$ b. $3.0805 \cdot 10^{12}$
   c. $4.6 \cdot 10^{-4}$ d. $9 \cdot 10^{-7}$

4. $\dfrac{6.0 \cdot 10^{24}\text{ kg}}{1.0 \cdot 10^{26}\text{ kg}} = \dfrac{6.0}{10^2} = 6/100 = 3/50$. The earth's mass is (about) 3/50 of Neptune's mass.

5. We need to divide to find out how many gold atoms "fit" into 99 grams of gold:

$\dfrac{9.9 \cdot 10^1\text{ g}}{3.3 \cdot 10^{-22}\text{ g}} = 3 \cdot 10^{23}$. There are about $3 \cdot 10^{23}$ gold atoms in 99 grams of gold.

Irrational Numbers

6.

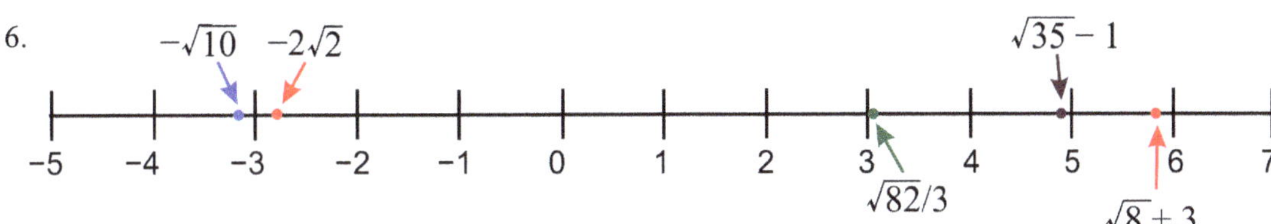

7.

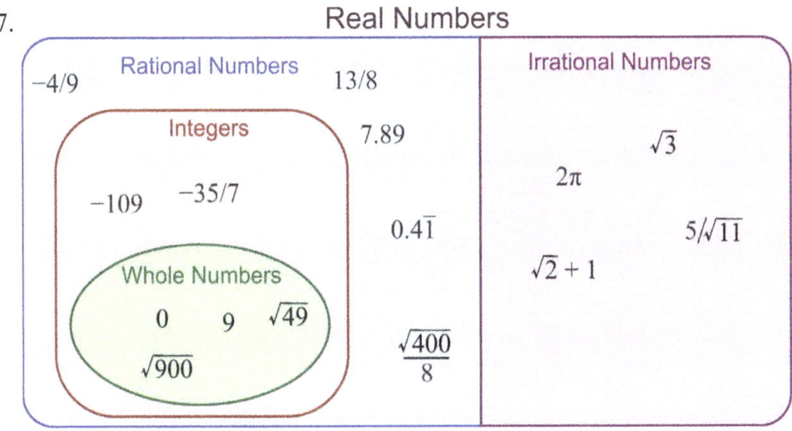

8. a. $x = \sqrt{54}$ or $x = -\sqrt{54}$ b. $n = 7$ or $n = -7$ c. $z = 4$

9. Let x = $0.\overline{71}$. Then $100x = 71.\overline{71}$. Subtracting those, we get:

$$
\begin{aligned}
100x &= 71.717171\ldots \\
-\ x &= 0.717171\ldots \\
\hline
99x &= 71 \\
x &= \underline{71/99}
\end{aligned}
$$

Geometry

10. Answers will vary. Check the student's answer. For example:

First, dilate triangle ABC from point A with scale factor 2/3.
Then translate it 8 units to the right.
Lastly, reflect it in the horizontal line $y = -1.5$.
(See the image on the right.)

But there are many possible answers. Here is another one.

First, reflect the triangle ABC in the horizontal line $y = -1.5$.
Then, translate it 8 units to the right.
Lastly, dilate it from point A" with scale factor 2/3.

Another one:
First, translate the triangle ABC 8 units to the right.
Then dilate it from point A' with scale factor 2/3.
Lastly, reflect it in the horizontal line $y = -1.5$.

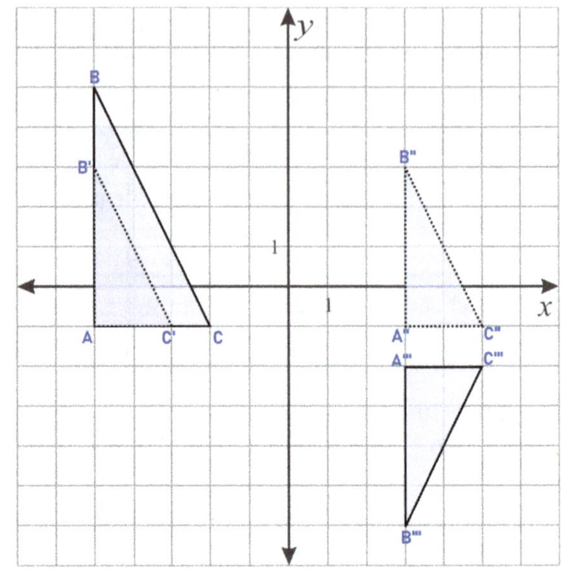

11. Answers will vary. Check the student's answer. For example:

First, rotate trapezoid ABCD 90° counterclockwise around the origin.
Then, dilate it from point C′ with scale factor 1/2.
(See the image on the right.)

Another way:

First, dilate trapezoid ABCD from point C with scale factor 1/2.
Then rotate it 90° counterclockwise around the origin.

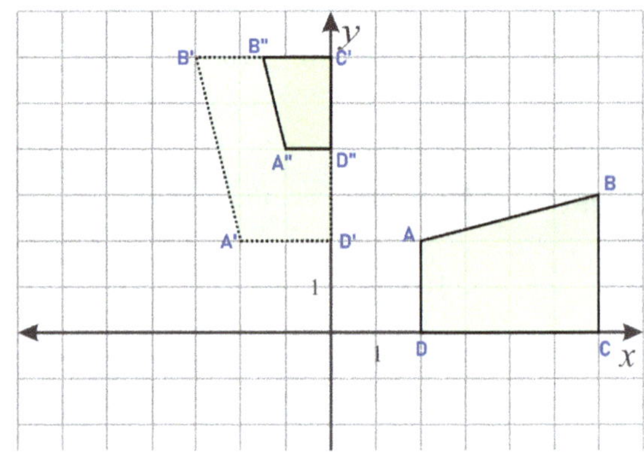

12. The bottom row of the table gives the coordinates after all the transformations.

Position	Vertices		
Original	A(−2, 5)	B(−5, 4)	C(−3, 0)
After reflection in y-axis	A'(2, 5)	B'(5, 4)	C'(3, 0)
After translation (2 units down, 1 to the left)	A"(1, 3)	B"(4, 2)	C"(2, −2)
After rotation (90° clockwise around the origin)	A'''(3, −1)	B'''(2, −4)	C'''(−2, −2)

13. Since the sum of the angles in a triangle is 180°, in triangle CDE, angle $a = 180° - 52° - 51° = 77°$.
Similarly, in triangle ABC, angle $b = 180° - 52° - 59° = 69°$.

Angles a, x, and b form a straight angle, so, $x = 180° - a - b$
$= 180° - 77° - 69° = \underline{34°}$.

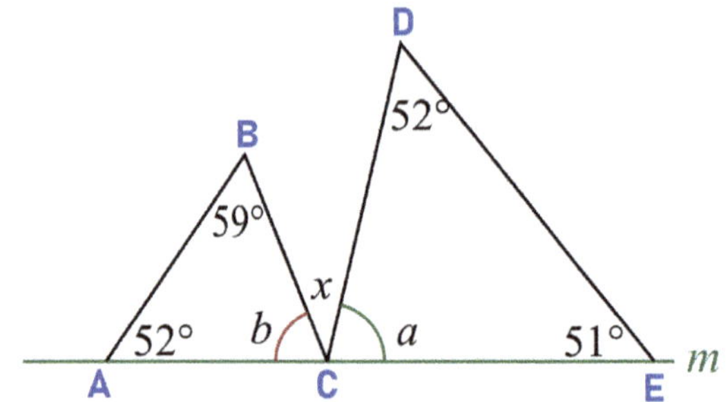

14. a. Angle BAC = 180 − (x + 26) = 154 − x. Now, the angle sum of triangle ABC is 180°, so, we can write an equation using that fact, and then solve for x:

$$\begin{aligned} \angle BAC + \angle ABC + \angle BCA &= 180 \\ 154 - x + x - 48 + x - 7 &= 180 \\ x + 99 &= 180 \\ x &= 81 \end{aligned}$$

b. Since m and n are parallel, angles y and BAC are corresponding angles, thus congruent. So, y = 154° − x = 154° − 81° = <u>73°</u>.

15. V = (4/3) · π (3.0 in)^3 · (2/3) ≈ 75.398 cubic inches. In cups, this is 5.2 cups or about 5 1/4 cups.

16. Let h be the height of the cup. Then, the volume is given by V = π (3.1 cm)^2 · h = 340 ml. This is an equation that we can use to solve for h. Since 1 ml = 1 cubic centimeter, the equation becomes π (3.1 cm)^2 · h = 340 cm^3, from which h = (340 cm^3)/(π · 3.1^2 cm^2) ≈ <u>11.3 cm</u>.

Linear Equations

17.

a.	b.
$$\begin{aligned} 10s + 8 &= 7s - 2(s - 5) \\ 10s + 8 &= 7s - 2s + 10 \\ 10s + 8 &= 5s + 10 \\ 5s &= 2 \\ s &= 2/5 \end{aligned}$$	$$\begin{aligned} 20 - 3(x + 4) &= 14 - 5x \\ 20 - 3x - 12 &= 14 - 5x \\ 8 - 3x &= 14 - 5x \\ 2x &= 6 \\ x &= 3 \end{aligned}$$

18.

a.	b.
$$\begin{aligned} \frac{2x - 3}{5} - x &= 2 \quad \Big\vert \cdot 5 \\ 2x - 3 - 5x &= 10 \\ -3x &= 13 \\ x &= -13/3 \end{aligned}$$	$$\begin{aligned} \frac{y - 3}{4} &= \frac{1 - y}{5} \quad \Big\vert \cdot 20 \\ 4(1 - y) &= 5(y - 3) \\ 4 - 4y &= 5y - 15 \\ 4 - 9y &= -15 \\ -9y &= -19 \\ y &= 19/9 \end{aligned}$$ or cross-multiply

19.

a.	b.	c.
$$\begin{aligned} 6x - 1 &= 6(x - 1) \\ 6x - 1 &= 6x - 6 \\ -1 &= -6 \end{aligned}$$ No solutions.	$$\begin{aligned} -5x + 1 &= 6(x - 1) - 5 \\ -5x + 1 &= 6x - 6 - 5 \\ -5x + 1 &= 6x - 11 \\ -11x &= -12 \\ x &= 12/11 \end{aligned}$$ One solution.	$$\begin{aligned} 6x - 12 &= 6(x - 2) \\ 6x - 12 &= 6x - 12 \\ 0 &= 0 \end{aligned}$$ An infinite number of solutions. Any value of x is a solution.

20. Let d be the amount of discount. The non-discounted blocks cost 3000($1.35) = $4,050. The discounted blocks cost $1500(1.35 - d)$. The total of these equals $5,775.

$$\begin{aligned} 4050 + 1500(1.35 - d) &= 5775 \\ 4050 + 2025 - 1500d &= 5775 \\ 6075 - 1500d &= 5775 \\ -1500d &= -300 \\ d &= 3/15 = 1/5 = 0.2 \end{aligned}$$

The discount was $0.20 per block. In other words, he paid $1.15 each for the 1500 blocks.

21. Let x be the first one of the four consecutive numbers. Then:

$$\begin{aligned} x + (x + 1) + (x + 2) + (x + 3) &= 2342 \\ 4x + 6 &= 2342 \\ 4x &= 2336 \\ x &= 584 \end{aligned}$$

The numbers are <u>584, 585, 586, and 587</u>.

22. Let p be the original price of the item. Then:

$$\begin{aligned} 1.06(0.73p) &= 34.82 \\ 0.7738p &= 34.82 \\ p &\approx 44.9987 \end{aligned}$$

The item cost $45.00 originally.

Functions

23. a. Because 3 is mapped to two different outputs: to 0 and to 3.
    b. The number 6 works. If you place either 3 or 9 there, then you will have the same input mapping to two distinct outputs, which would make it not a function.

24. a. Farm B's pricing system is a linear function. $C = 6.25w$.
    b. For Farm A, the rate of change is $(21.5 - 15)/(3 - 2) = 6.5$, or $6.50 per kg.
    For Farm B, the change of rate is 6.25, or $6.25 per kg.
    c. At Farm A, 4 kg will cost about $27, and at Farm B, $25. So, Farm B has the better deal.
    For 7 kg, Farm B charges you $40 and Farm B $43.75, so, Farm A has the better deal.

25. a. $10.
    b. That there is an initial fee of $10 just to get to go riding.
    c. $1 per minute.
    d. Horse riding will cost you $1 per minute, on top of the $10 initial fee.
    e. cost = $10 + t$, where t is the number of minutes you will go riding.

26. a. From $x = 0$ to $x = 4$: linear and increasing
From $x = 4$ to $x = 8$: nonlinear and decreasing
From $x = 8$ to $x = 11$: linear and decreasing
From $x = 11$ to $x = 15$: nonlinear and increasing

b. 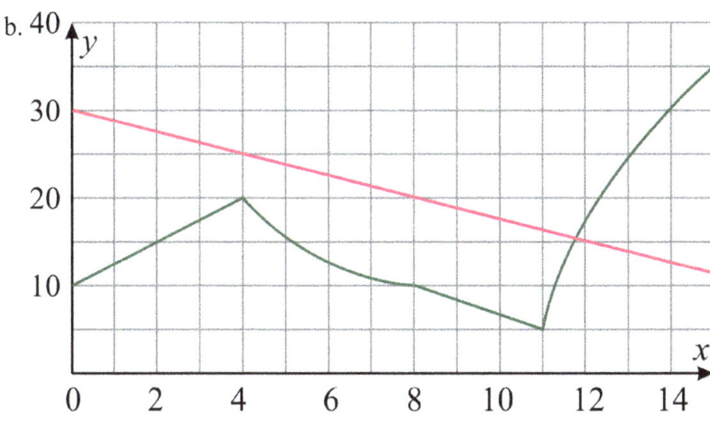

c. $y = -(5/4)x + 30$

d. For the function in green: the rate of change is −5/3. For the function in red, it is −5/4.

Graphing Linear Equations

27. a. $y = (-2/3)x + 4$
b. $y = -3$
c. $y = 5x - 25$

28. a. Fridge 1 (120 kWh versus 100 kWh). It consumes 20 kWh more than Fridge 2.
b. Fridge 1: $E = 40t$. Fridge 2: $E = (100/3)t$. It is also acceptable to write it with a rounded decimal, as $E = 33.3t$.

c. 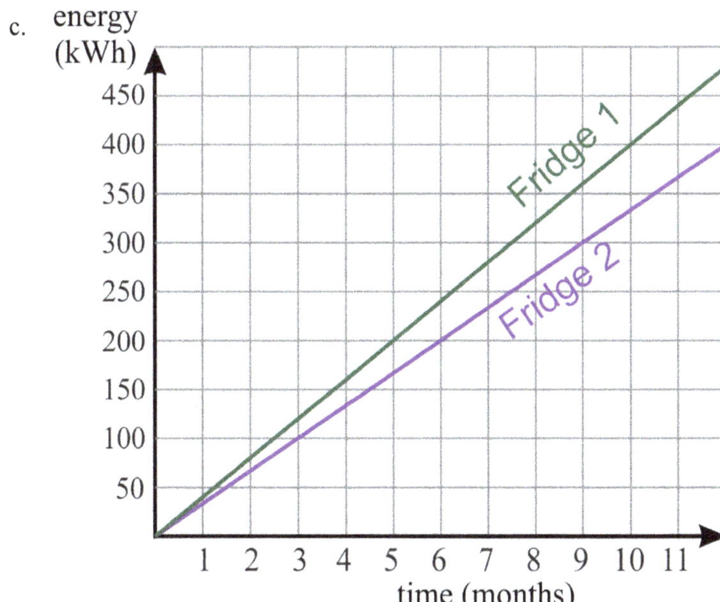

29.

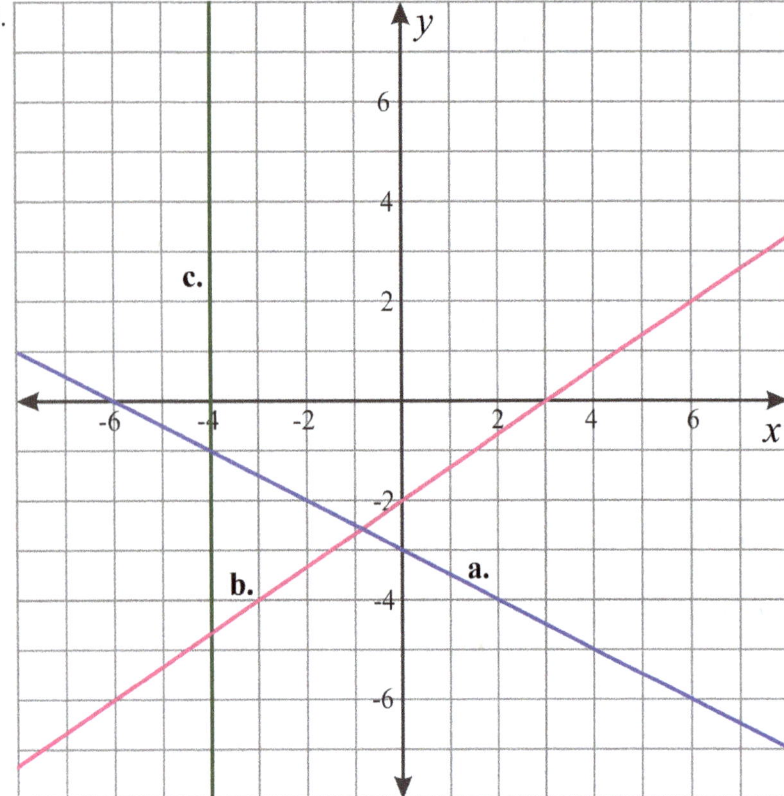

30. The slope of this line can be calculated using the two given points. It is $(-6 - 14)/(-7 - 3) = -20/(-10) = 2$. The equation of this line is therefore of the form $y = 2x + b$. Substituting (3, 14) into it, we get $14 = 2(3) + b$, from which $b = 8$. So, the equation is $y = 2x + 8$. Since point $(a, 2)$ is on this line, let's substitute those values into the equation of the line:

$$\begin{aligned} 2 &= 2a + 8 \\ -6 &= 2a \\ a &= -3 \end{aligned}$$

So, $\underline{a = -3}$. There are also other ways to arrive to the final answer, such as using the formula for the slope.

The Pythagorean Theorem

31. Using the Pythagorean Theorem, we get:

a. $\begin{aligned} r^2 + 17.5^2 &= 26.6^2 \\ r^2 &= 26.6^2 - 17.5^2 \\ r^2 &= 401.31 \\ r &= \sqrt{401.31} \approx 20.0 \end{aligned}$ We ignore the negative root since this is a length of a side. The unknown side measures 20.0 units.	b. $\begin{aligned} x^2 + x^2 &= (\sqrt{70})^2 \\ 2x^2 &= 70 \\ x^2 &= 35 \\ x &= \sqrt{35} \end{aligned}$ We ignore the negative root since this is a length of a side. The unknown side measures $\sqrt{35}$ units.

32. The rafter, the height of 1 ft 10 in, and half of the 6 ft 8 in span form a right triangle. In this triangle, using inches instead of feet and inches, the two legs measure 22 in and 40 in. Now, let r be the length of the rafter. According to the Pythagorean Theorem:

$$\begin{aligned} r^2 &= 22^2 + 40^2 \\ r^2 &= 2{,}084 \\ r &= \sqrt{2{,}084} \approx 45.651 \end{aligned}$$

The decimal portion, 0.651 inches, can be converted into 16th parts of an inch this way. Let x be the number of 16th parts of an inch that equals 0.651. Then, $x/16 = 0.651$, from which $x = 0.651(16) = 10.416$.
So, the rafter measures 3 ft 9 10/16 in.

33. a. To find the height, we will use the right triangle ABC. First, we need to find the length of the diagonal of the bottom square (d). From the Pythagorean Theorem:

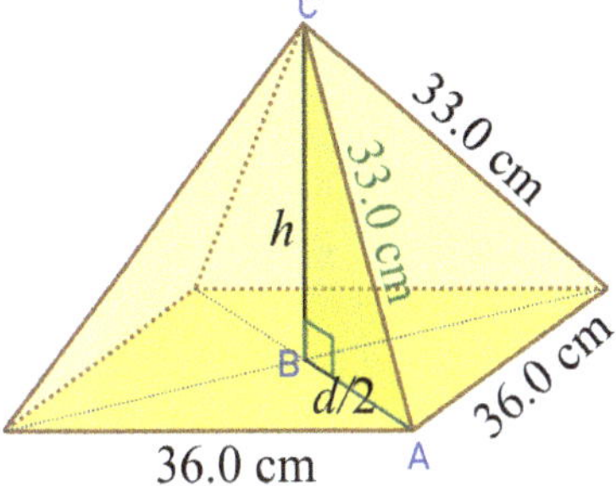

$$\begin{aligned} d^2 &= 36.0^2 + 36.0^2 \\ d^2 &= 2592 \\ d &= \sqrt{2592} \end{aligned}$$

Next, we apply the Pythagorean Theorem to triangle ABC. Note that one of its legs is $d/2$ (half of the diagonal), the other leg is the height of the pyramid (h), and the hypotenuse is the 33-cm edge of the pyramid.

$$\begin{aligned} h^2 + (\sqrt{2592}/2)^2 &= 33.0^2 \\ h^2 + 2592/4 &= 1{,}089 \\ h^2 &= 1{,}089 - 648 \\ h &= \sqrt{441} = 21 \end{aligned}$$

The height of the pyramid is 21.0 cm.

b. The volume is V = 36.0 cm · 36.0 cm · 21.0 cm / 3 = 9,072 cm^3.

Systems of Linear Equations

34.

a. $\begin{cases} 2x - 3y = 8 \\ 3x + 4y = -5 \end{cases} \quad \left\vert \begin{array}{l} \cdot 3 \\ \cdot (-2) \end{array} \right.$ $\downarrow$ $\begin{array}{rrcl} & 6x - 9y & = & 24 \\ + & -6x - 8y & = & 10 \\ \hline & -17y & = & 34 \\ & y & = & -2 \end{array}$ Substituting $y = -2$ in the first equation, we get: $\begin{aligned} 2x - 3(-2) &= 8 \\ 2x + 6 &= 8 \\ 2x &= 2 \\ x &= 1 \end{aligned}$ Solution: (1, −2)	b. $\begin{cases} -x = 4(y + 5) \\ 2x = -12y - 10 \end{cases}$ Solving for x from the top equation, we get that $x = -4(y + 5)$ which simplifies to $-4y - 20$. Now, substituting that for x in the bottom equation, we get: $\begin{aligned} 2(-4y - 20) &= -12y - 10 \\ -8y - 40 &= -12y - 10 \\ 4y - 40 &= -10 \\ 4y &= 30 \\ y &= 30/4 = 15/2 \end{aligned}$ Substituting $y = 15/2$ in the first equation, we get: $\begin{aligned} -x &= 4(15/2 + 5) \\ -x &= 4(25/2) \\ -x &= 50 \\ x &= -50 \end{aligned}$ Solution: (−50, 15/2)

35. a. No solutions. b. One solution. c. An infinite number of solutions.

36. Let x be the number of tables that seat 4, and y be the number of tables that seat 6.

We can write this system of equations: $\begin{cases} x + y = 106 \\ 4x + 6y = 500 \end{cases}$

Solving for y from the top equation, we get $y = 106 - x$. Substituting that in the bottom equation, we get:

$$\begin{aligned} 4x + 6(106 - x) &= 500 \\ 4x + 636 - 6x &= 500 \\ 636 - 2x &= 500 \\ -2x &= -136 \\ x &= 68 \end{aligned}$$

Then, $y = 106 - x = 106 - 68 = 38$. The restaurant has <u>68 tables that seat 4, and 38 tables that seat 6</u>.

37. Let G be Greta's age and S be Susan's age. Then: $\begin{cases} G + 10 = (3/4)(S + 10) \\ G + S = 127 \end{cases}$

From the bottom equation, we can solve that G = 127 − S. Substituting that in the top equation, we get:

$$\begin{aligned} 127 - S + 10 &= (3/4)(S + 10) \\ 137 - S &= (3/4)S + 7.5 && \cdot 4 \\ 548 - 4S &= 3S + 30 && + 4S \\ 548 &= 7S + 30 && - 30 \\ 518 &= 7S && \div 7 \\ S &= 74 \end{aligned}$$

Then, G = 127 − 74 = 53. <u>Greta is 53 and Susan is 74</u>.

38. a. Nonlinear and decreasing association. b. No association. c. Linear and increasing association.

39. There is no association between the variables. For each age group, there is about an equal number of people who exercise and who do not exercise. (In other words, in each age group the relative frequencies for "Exercises" and "Does not exercise" would be close to 50%.).

40. a. 6 b. 24 c. 3

41. a. Answers will vary. Check the student's answer. For example:

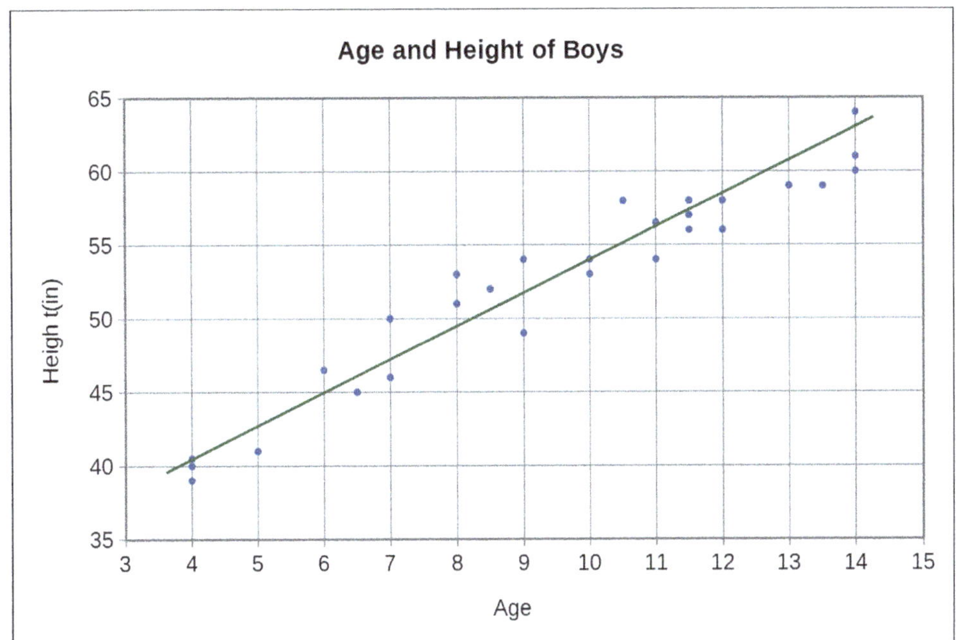

b. Answers will vary. Check the student's answer. The line above goes through (6, 45), and (10, 54). Therefore, its slope is 9/4 = 2.25, and its equation is of the form $y = 2.25x + b$. Substituting (10, 54) into this allows us to solve for b: $54 = 2.25(10) + b$, from which $b = 31.5$. So, the equation is $y = 2.25x + 31.5$.

c. It means that each 1-year increment of age is associated with a 2.1-inch increment in height. In other words, boys tend to grow 2.1 inches per year.

d. The y-intercept of 32.6 inches means that this equation predicts a newborn baby to be 32.6 inches tall. However, we know newborns are not that tall; they are typically between 18-22 inches tall. They grow very fast during the first year. Then from about age 2 onward, the growth follows a linear pattern fairly closely. This shows us that we cannot extrapolate backwards all the way to zero years using this data and this equation.

e. We solve the equation $50.5 = 2.1x + 32.6$ for x:

$$\begin{aligned} 50.5 &= 2.1x + 32.6 \\ 50.5 &= 2.1x + 32.6 \\ 17.9 &= 2.1x \\ x &= 8.52 \end{aligned}$$

The equation predicts the age of about 8.5 years for a boy that is 50.5 inches tall.

www.ingramcontent.com/pod-product-compliance
Lightning Source LLC
Chambersburg PA
CBHW080549220326
41599CB00032B/6420

* 9 7 8 1 9 5 4 3 5 8 5 7 7 *